P9-ASJ-302

Magicolor Studios

About the Author

LYNNELL HANCOCK is an assistant professor at the Columbia School of Journalism, where she served as director of the Prudential Fellowship for Children and the News, a program dedicated to improving media coverage of children's issues. She has been a writer and editor at *Newsweek,* the *New York Daily News,* and the *Village Voice,* and now contributes to *U.S. News & World Report* and *Parenting.* She lives in Montclair, New Jersey.

Hands to Work

Hands to Work

Three Women Navigate the New World of Welfare Deadlines and Work Rules

to

Work

LynNell Hancock

Perennial

An Imprint of HarperCollinsPublishers

A hardcover edition of this book was published in 2002 by William Morrow, an imprint of HarperCollins Publishers.

HarperCollins books may be purchased for educational, business, or sales promotional use. For information please write: Special Markets Department, HarperCollins Publishers Inc., 10 East 53rd Street, New York, NY 10022.

First Perennial edition published 2003.

Designed by Kate Nichols

The Library of Congress has catalogued the hardcover edition as follows:
Hancock, LynNell.
    Hands to work : the stories of three families racing the welfare clock / by LynNell Hancock.—1st ed.
      p. cm.
    Includes bibliographical references and index.
    ISBN 0-688-17388-8 (alk. paper)
    1. Welfare recipients—New York (State)—New York—Case studies. 2. Poor women—New York (State)—New York—Case studies. 3. Public welfare—New York (State)—New York—Case studies. 4. Welfare recipients—Employment—New York (State)—New York—Case studies. I. Title.

HV99.N59 H36 2002
361.6'8'09747275—dc21
                                      2001031730

ISBN 0-06-051216-4 (pbk.)

04   05   06   07   ❖/RRD   10   9   8   7   6   5   4   3   2

To Eugene Hancock, my dad and my first teacher

Work without hope draws nectar in a sieve,

And hope without an object cannot live.

—SAMUEL TAYLOR COLERIDGE, 1828

Contents

Hands to Work

Introduction

Hands to Work

I grew up with a little girl's worry that everything attached to my otherwise winsome Iowa world might disappear in a whisper. The hulking dusk-plum Oldsmobile with the chrome fins in our driveway. My Ginny doll with her store-bought shirtwaist dress. Our behemoth Zenith television set. My red Schwinn, shorn of those childish training wheels. These were the fret-free fifties, yet the Great Depression hung over our kitchen table in the farming town of Independence like an imminent midwestern twister. It was best not to get too enamored with any material things. They could be snatched away by circumstance, by misfortune. Where would we be if my dad lost his Methodist minister's job?

To drive these anxieties home, my mother served a steady diet of "Depression food"—oatmeal, hash, recycled leftovers. She posted warning signs to ward off the specter of deprivation that lingered over our porch like an exotic visitor, tempting, murmuring. Taped above our yellow metal table was a picture of a starving girl in India, empty bowl extended, reminding us to clean our plates. A cross-stitched sampler done up in salmon and green embroidery threads hung over the sink, admonishing us to complete our chores:

HANDS TO WORK, HEARTS TO GOD.

In the world of a seven-year-old, work was endowed with almost mystical qualities. Doing a good job was almost as beneficial as going to church three days a week, or reciting all the books of the Bible in perfect order. I translated my mother's admonitions in the most literal ways. If I could just keep making my bed, cleaning my plate, busying my hands with needlework, not only would the starving girl in India be kept at bay, but salvation would be ever nearer. It was the biggest work bonus of all. God smiled down on the perpetually busy. He didn't much care for those who doodled away their days in the backyard playing jacks. Lazy people in our house were called "work brittle." It was a precarious way to live. It made you vulnerable to strong gusts, to failure. One of my dad's Ozark uncles was work brittle, and he ended up losing the family farm.

Poverty didn't fit very well into this picture, and yet there it was—an ever naive possibility. Our favorite family board game was Life, where to lose meant to end up on a one-way detour to the "Poor House." My dad's stories of growing up poor in southern Illinois contributed to this morbid obsession. He was part of a big family—six boys, two girls. The boys all got one pair of overalls and one pair of shoes on Easter each year. In the warm months they went barefoot to conserve their soles. They would stick pins into the calluses on their feet, winning this game by pricking the deepest.

Grandpa Pat Hancock worked as a clerk for the Chicago & Eastern Illinois Railroad, filling record books with his elegant script chronicling the comings and goings of trains in the roundhouse. Grandma had to quit her job teaching school in Buggy Ridge, Missouri, to raise this family. (Women in 1910 were not allowed to teach public school and be married at the same time.) The couple set up house in a modest white clapboard home complete with a wood-burning stove and privy out back.

But then Grandpa lost his job during the Depression, like just about everybody else in West Frankfurt, Illinois. The sons picked blackberries in the woods and sold them door to door to the few people who still had work. Grandma and her daughters took in other folks' washing. The plagues of the thirties came with random regularity. Small pox. Blood poisoning. Tuberculosis. Two babies died at birth. The Hancock family was never quite the same. Work couldn't save them.

These were proud people, Methodists of the no-dancing variety, with values and traditions straight out of New England's preindustrialized Puritan homesteads. Work, at least for Grandma, was not just pedantic toil that put food on the table. It was a form of worship. It was a way to create yourself anew, as the ancient Greeks believed, as the philosopher Hegel codified centuries later. It was a way to contribute to your own salvation. There was little choice in the matter.

Most important, you "did for yourself." Nobody else could take over such a weighty task for you. Accepting something for nothing was not only distasteful, it was sinful. When autonomy was impossible, for monumental economic reasons, life became a calamity that simple hard work could not put right. My dad overheard his father say he would never take those Depression rations. He never accepted charity in his life, and he was not about to start. Grandpa felt personally responsible, a great failure, because of his family's predicament. Pat Hancock would never have thought to blame the nation's staggering financial, industrial, and political collapse for his own misfortune.

Soon even Grandpa had no choice. There wasn't enough piecemeal work to feed his family in the late thirties. He would walk blocks ahead of his teenage son, my dad, who would pull the free rations in a wagon. Finally, Grandpa turned for help to the nation's first massive federal welfare program, designed by Franklin Delano Roosevelt. It wasn't called welfare then, it was called relief, and just about everybody in town was signing up to get some. Grandpa helped build a public school with the Civil Works Administration—an unprecedented federal work relief experiment that employed more than four million men. His nephews, neighbors, and cousins got work at the Civilian Conservation Corps camps. Work, purposeful work, not the dole, became the cornerstone of Depression-era welfare. Men were paid subsistence wages for their efforts. Their labor built many valued roads, utilities, hospitals, schools.

Relief kept just enough food around to ward off malnutrition. Grandma filled her brood with cooked oats and Bible stories. For special occasions she'd make Depression Cake, otherwise known as Empty Pockets Dessert—a miraculous spice confection held together without the aid of hard-to-get eggs or butter. The family made do with a lot of help from the government, a source of deep shame for Grandpa. "I should be able to take care of my own," he'd tell his wife in low tones so the children wouldn't hear. He had incorporated lessons learned a long time earlier, that charity breeds idleness, that handouts corrupt a person's will to work.

Despite his father's reticence, my dad wasn't ashamed of his family's poverty. Back then, it didn't set him apart. He shared the same fate as many of his pitching-penny buddies. He was nobody special, the middle son of a family living in a small, southern town with equal opportunity joblessness. "Nearly everyone we knew was in the same predicament," he would say. Unemployment hovered around 80 percent in their county. The racially segregated small towns of southern Illinois were not marred by the disparate income gaps of the East Coast and urban areas where the more common "dole-chiseler" sentiments flourished. In the plains states, so many hardworking men— white fathers—were down on their luck that there was a general sense of compassion, rather than contempt, for those who had to sign up with FDR's New Deal projects to survive. As a result, my dad grew up with a sense that the federal government was obliged to provide for anyone who was down and out. It was the Christian thing to do. It was charitable. All people had a basic human right to eat and have shelter, no matter what happened to them. Enormous programs—the Works Progress Administration, Social Security, the Federal Emergency Relief Act—prevented mass starvation, and probably mass disorder. For years to come, Dad preached from his Methodist pulpit about the value of social generosity.

Long before the Depression and long after, these dueling sentiments—my grandfather's shame and my father's dutiful compassion— have fed the ambivalence that Americans feel toward their poor, and toward federal policies aimed to help them. Government either degrades one's dignity and corrupts the will to work or rescues entire families from poverty. It either creates poverty by the weight of its own stifling generosity or attempts to return some dignity to those who have little left. This philosophical tug and pull, this profound incongruity, is one reason American welfare has always been so meager compared to Europe's more expansive social programs. Americans have never come to terms with it.

Still, if the Depression taught Americans anything, it was that citizens need some assurance that a Wall Street collapse will never again take food out of their children's mouths. If big government programs can provide that cushion, then so be it. Nearly everyone in America was in some way scathed by the Depression. It wasn't a scourge that landed selectively on inner-city blacks, or on rural mountain folks. Nearly everyone for a brief blip in history was on board with a federal solution, particularly if work was mixed into the equation.

The modern welfare state evolved rather naturally at the end of the Depression with Roosevelt's Social Security Act of 1935. One tier, for Social Security and unemployment insurance, survived untainted by the stigma of welfare. The other tier, general assistance for the blind, disabled, and otherwise unemployable, became more problematic. The most controversial program, called Aid to Dependent Children (later Aid to Families with Dependent Children), began as a relatively meager allowance for war widows with young children. In the 1960s AFDC applicants began flooding into the nation's welfare centers, encouraged by the federal War on Poverty and by the civil rights movement. More than three million families were receiving government aid by the decade's end, a fourfold increase. For the first time many of them were black, Hispanic, urban, single, divorced mothers. Welfare had made a full turn from a social service to an entitlement—for the minority poor.

From then on, few American presidents failed to declare welfare reform a top priority. First, cash aid was reduced. Then fraud detection was beefed up. Work requirements were always required in one form or another. Reforms in the eighties did little to cut down the number of recipients. By 1995 disparagement of welfare was so accepted that members of Congress were calling recipients "alligators" and "wolves" on the floor of the House of Representatives. The poor presented a shameful scar on America's patina of success as the richest nation in the world. The fact that one in five American children lived in poverty made a mockery of the nation's role as a global powerhouse.

The Democratic administration decided that the prosperous nineties was the time to put an end to it. But this time the cry was not to end poverty, but to end "welfare as we know it."

When President Bill Clinton lowered his pen in the White House Rose Garden on a steamy August 22, 1996, he signed perhaps the most far-reaching bill of his stormy tenure. It certainly came to have the most ironic of names, considering this U.S. president would become well known for his own grave flaws in judgment. The Personal Responsibility and Work Opportunity Reconciliation Act reconstituted federal welfare under the heady slogan that "a hand up" is better than a "handout." This jingle struck the right chord with the average middle-class American, who was likely to be working one and a half jobs for fewer inflation-adjusted dollars and benefits than in the past.

It was another trademark political move for President Clinton, ensuring his place in history as master of the deal, catapulting him toward a reelection sweep three months later. Borrowing a phrase from the welfare state's most conservative opponents, the Democratic president announced that this bill would make welfare "a second chance, not a way of life."

The reform act was not a piece of legislation meant to tame class warfare or balance the federal budget. Money was not the point. Welfare spending on families with children represented less than 1 percent of the national budget. Nor was social order an immediate concern. There were no raging racial riots to quell, as there once had been in Watts, in Birmingham, in Newark, in Selma. Instead, the new law was driven by old ideas, reflections of my grandfather's philosophy that welfare perpetuates the poverty it is meant to relieve. It was a statement on our nation's values. Once again, poor individuals were ultimately held responsible for their own deprivation. With this welfare document, the children of Depression parents were passing judgment. As the historian Michael B. Katz wrote, "Welfare became the lightning rod for Americans' anxieties over their work, incomes, family and futures." Hardworking baby boomers were resentful that some segments of the population were not putting in toil time for their benefit checks.

The new social contract would be carried out using tough, paternalistic love. State governments would set the behavior standard for those living at subsistence or less than subsistence levels, and then enforce it—some with generous incentives, others with punitive measures. For the first time in American social welfare history, strict deadlines were enacted. No family could receive federal cash assistance for longer than five years, the limit for an entire lifetime. Single adults were restricted to two years. States were given far more latitude to spend chunks of money earmarked for welfare. The hope was that local governments would find innovative ways to invest the money in child care, job programs, training sites, or employment supplements— whatever made sense to support their own populations.

Work became the policy's cornerstone. After two years' time at least half of each state's welfare population would be required to "work," by that word's broadest definition. If the recipient couldn't find private-sector work, then government would provide something— work that was very different in nature from the Depression-era projects. In urban areas like New York City people in need were handed brooms and sponges instead of bricks and shovels. Laborers didn't receive wages per se, but a package of benefits. Checks could be pulled for failure to show up at the labor site. Thousands of the country's poorest citizens cleaned parks, swept streets, and scrubbed municipal floors. No airports, murals, or roads were constructed.

I wondered how my grandfather would have survived.

This breathtaking social revolution was begun with few roadmaps, and fewer guiding studies. New York's Mayor Rudolph Giuliani referred to it as a Montesquieu-era social contract between government and the poor—we give, you return the favor in kind. The results in raw statistics were astonishing. The number of people receiving government aid fell by 50 percent nationwide in the first four years of the reform effort. Work in exchange for aid was a formula that sat well with most taxpayers, as it had during the Depression. Yet the world of work was a very different place in the nineties than it was before World War II. Technology had changed not only the modes of production but the training needed to earn enough for basic family staples. More than 80 percent of the current jobs required a high school diploma. The Department of Labor reported that the fastest-growing occupations, in areas like computer and data processing, required college degrees.

Still, the bill's underlying assumption was that anyone who wanted to work in the prosperous twenty-first century would find it, regardless of education level. The vast majority of Americans wanted desperately to believe that work in its purest form—not training, not education, not more generous aid packages—would overcome these dire predictions. Work would be the poor's salvation—in the here and now and the hereafter. "People on welfare ought to work, work, work, because it is good for the soul," said Colorado's Senator William Armstrong.

The New York City welfare commissioner put it this way: "Work is one's gift to others," said Jason Turner. "When you sever that relationship, you're doing more than just harm to yourself economically. You're doing spiritual harm."

It seemed that my grandfather's Puritan ethic, and my own primitive understanding of the magical powers of keeping busy, had overtaken public policy. Work would end poverty. Or was ending poverty the point?

This new welfare world is an emerging, untested social experiment—one that has the potential to define what kind of nation we want to be, what kind of government we think is most fair. It's a political story. It's an economic story. It's a story about social reinvention. But in the end it is simply a human saga. It is about ordinary Americans trying to make a life for themselves, caught by an accident of timing in the wake of a social experiment meant to change the course of their lives. Five years after the clock began ticking on lifetime welfare limits, much of the public debate still swirled around the most rudimentary subtraction—the shocking number of recipients who dropped off the dole. Seven million people. The number was used as the most important measuring stick of welfare reform's success. So little of that discussion was about these people themselves, what happened to them, how their fates might affect all our lives.

This book tells the stories of three of those families, surviving poverty in neighboring Bronx streets, as they found their own way through new rules and deadlines in the world of welfare. All three petitioned the government for aid in the midst of this bureaucratic and political upheaval. Their reasons for seeking help were as divergent as the individuals themselves. They all imagined far different futures for themselves and their families as they careened toward the first-ever lifetime limits on benefits. Their stories cannot hope to represent the infinite experiences of an entire welfare community. They simply provide a window into a few of the human stories behind the policy. In an era of skyrocketing stock market leaps, when the gap between wealthy and poor is widening into an epic chasm, these families must either blend comfortably into the fabric of America's prosperity or dissolve from view.

Alina Zukina was a young, ambitious Russian refugee whose family had fled anti-Semitism in their native Moldova only to find new forms of discrimination in New York City. The Zukinas faced the wall of mistrust that meets many immigrants. At the same time they faced the often illogical hurdles that had been carefully constructed for those seeking public aid. Alina worked the city's workfare program for four years in exchange for her small benefit check, all the while studying at night to fulfill her dream of becoming a doctor. In a strange twist of fate, Alina encountered the same dilemma in America that had forced her to leave the country she loved years earlier. With just one semester left before graduation, Alina had to choose between her degree and her welfare check.

Christine Rivera came from a long line of proud Puerto Rican migrants to the barrios of the Bronx. A mother of four, a onetime medical assistant, Christine was well on her way to self-sufficiency at the millennium despite the cornucopia of untreated trauma in her life. In many ways her life was at its peak two years earlier, when two unfortunate events collided: her stubborn heroin addiction and the new world of welfare. Within a few months' time, instead of approaching the brink of independence, Christine found herself facing a long life in lockup and the possible loss of her children.

Brenda Fields was a single mother of two, and by all rights, a model candidate for the new welfare reformers. Energetic, resilient, eager to work, Brenda could think of nothing more appealing than to shed the humiliating "welfare recipient" moniker and make something of herself. She believed in the message: work would liberate. Brenda set about to work around the new rules, trying to overcome her lack of a college degree and her limited work experience. But there was never enough time. Never enough money. Her journey takes us through the homeless shelter, into the welfare centers, into the world of the working poor, and back again. Through it all, Brenda never stopped believing that it would all work out in the end.

I did not choose these women based on any scientific calculation. I chose them simply because I was drawn to their stories, and they were generous enough to let me invade every facet of their lives for three years and more. One happens to be white, one Puerto Rican, one African American. One is an immigrant, one a foster child, one a third-generation Bronxite. One is bold, one is shy, one is troubled. One has a strong family network, the other two have forged their own makeshift families from a potpourri of friends and lovers. One story takes us to the former Soviet Union, one to the world of the working poor, the third to the rawest life on the streets of the South Bronx, one of the nation's most prickly pockets of urban poverty. Their shared borough poses two daunting challenges to the success of welfare reform: high crime and high unemployment. As a whole, these women's tales are profiles in courage, bad judgment, worse luck, and ultimately, resilience.

Their journeys unfold over a four-year period against the backdrop of increasingly conservative welfare policies in the nation's most liberal city. The architect of welfare reform in this brash city was its famously brash mayor and his ultra-conservative welfare commissioner. Mayor Rudolph Giuliani made welfare reform the linchpin of his mayoral tenure as he worked to turn "the welfare capital of the nation into the workfare capital of the nation." Jason Turner made his mark in the welfare world by designing the landmark Wisconsin reforms, regulations that "put work first," he said, not checks first. Both believed deeply in their task at hand. Both set radical goals for the system—to transform welfare centers into job centers, to put everyone to work, and to end welfare by the year 2000.

The women in this book must endure beyond that deadline, long after they have disappeared from the rolls of welfare and the minds of politicians.

Part I

Journeys

The work of the world is common as mud.

Botched, it smears the hands, crumbles to dust.

But the thing worth doing well done

has a shape that satisfies, clean and evident.

Greek amphoras for wine or oil,

Hopi vases that held corn, are put in museums

but you know they were made to be used.

The pitcher cries for water to carry

and a person for work that is real.

—MARGE PIERCY, FROM "TO BE OF USE," IN *Circles on
the Water: Selected Poems of Marge Piercy*
(NEW YORK, 1982)

Brenda

Papers, Papers, Papers

Two worn canvas suitcases fell to the Brooklyn sidewalk with a defeated clatter. Brenda Lee Fields had been carrying her whole life in those bags for months—bags heavy with papers, clothes, papers, forks and spoons, papers, cigarettes, can openers, papers. Inside was a smudged certificate announcing Brenda's birth in a Queens hospital in March 1960; her class of '79 diploma from Fashion Institute High School; paperwork on her three-year-old son, Tyjahwon, and seventeen-year-old daughter, Loreal; a transcript for one semester of community college studies; and a letter from her Bronx landlord saying he'd sold her apartment out from under her. All of it was proof of her very existence, her current crisis, neatly folded inside an old Avon cosmetics bag. Somehow, none of the documentation was ever quite good enough.

It was nearly midnight in an unfamiliar part of Brooklyn, and this American family felt less like citizens than refugees, swept up in a foreign era of welfare reform that would determine the course of their lives. Fort Greene, once an important jumping-off point for the Civil War's Underground Railroad, was now home to an ethnic mix of working-class immigrants and Wall Street bankers. Ty, an unflappable toddler, buckled under the weight of his bookbag stuffed with clothes, a Thomas the Train toy, and a worn-out "Cat in the Hat" book. He stood before the classical structure looming above him, another assessment center that would serve as a temporary home and determine his short-term future.

"My house?" he asked his mom. "Is it a monster?" To the cranky, bleary-eyed three-year-old, the once-abandoned hospital building looked haunted. Loreal, a senior at the well-regarded Harry S. Truman High School in the North Bronx, just rolled her eyes. To a seventeen-year-old high school senior, the place merely represented another numbing commute. She tried to cipher how she could make it to her 8:00 A.M. class that was now ten miles away, at least. The regal teen was just months away from graduating. She had plans for college, for a life on her own. She'd missed a lot of school already this year in all the upheaval. Her immediate future was fading before her.

"Come on, let's get on in here," Brenda told her exhausted brood. She figured this relocation to a temporary dormitory in February 1997 was some new form of calculated humiliation. She still had nineteen dollars in her pocket from yesterday's emergency check. She still had her kids. But her proud composure was slipping. Her beautiful daughter's quiet disappointment hurt, deeply. "This city wants me to give up," Brenda thought as she calculated how long the three of them could make it on the few dollars in her pocket. Between sleeping on city agency floors and shuffling back and forth to temporary beds, she had missed at least three appointments to sign up for public assistance. The city was not making this easy.

Brenda contemplated the decrepit Auburn Assessment Center, a neglected former hospital turned makeshift city shelter for those with nowhere else to go. Here in the heart of the Walt Whitman Housing

Project was where hundreds of people were sent to wait for days, for weeks, while the city decided whether their circumstances were dire enough to qualify for city-run shelters. And qualifying as "dire enough" was becoming more difficult. At Auburn, rodents rushed along the lobby baseboards, leaving holes so large that wires bulged through like electrified hernias. Its shabby rooms were filled with families like Brenda's, people who had tried and failed to convince the increasingly skeptical city that they were indeed homeless.

Three years earlier Rudolph Giuliani had been sworn in as mayor of New York City pledging to "end welfare by the end of this century, completely." It was a quiet vow that would later become a hallmark of his administration. One important strategy in his new war on welfare was to combat fraud, making it much tougher to enter the system in the first place. To stem the flow of people seeking shelter, a timeworn first step onto the public dole, the mission of the Emergency Assistance Unit was changed from "assistance" to "fraud investigation." Under the new rules, applicants were discouraged from seeking shelter if they had relatives or friends with apartments anywhere in the region. Such an appeal would constitute deceit in the eyes of the investigators. The city would no longer be a sugar daddy, Giuliani reasoned. It would not offer simple sanctuary, but rather an encouraging segue into the work world. Government was not in business to pamper souls but to build hardworking citizens. Families that had once been considered eligible, pre-Giuliani, were now being routinely rejected. A few returned to viable apartments, but many were sent back to volatile homes, or dangerously crowded conditions.

City Hall and a large segment of the taxpaying public seemed pleased enough with the results. The nation's largest welfare rolls had dropped more than four hundred thousand names by 1997—the entire population of Buffalo, the mayor was fond of saying—just one year after the federal government passed its own reform laws. No city department was charged with following the paths of these welfare refugees after they vanished. Giuliani believed that tracking the poor was akin to playing Big Brother. Without hard numbers, the mayor could claim success. He pointed to low crime rates and empty beds in homeless shelters as proof that the former dependents were finding work and a new sense of purpose. The city's public advocate countered with a report that found twenty-four hundred people turned away unfairly from homeless shelters in a four-month period, even while nearly four hundred beds went unused. The busiest homeless processing outpost in the city had already collected some unwanted headlines. The *New York Times* ran several stories reporting that more than five hundred people per night were sleeping in the Emergency Assistance Unit in the Bronx. Adults and children were stretched out on chairs, on floors, eating in unsanitary conditions. A state appeals court later found conditions there "intolerable" and ordered the city to pay thousands of homeless families $2 million in damages.

One way to halt this traffic of needy people was to change the rules, to make it harder to qualify for shelter—and hence the welfare rolls. That's what happened in the summer of 1996, a few months before Brenda tested the waters. The city decided it would begin turning away people who told officials they had relatives with apartments anywhere in the New York area or beyond. Overnight two-thirds of the center's applicants were deemed ineligible. The first wave of rejects camped out behind the squat brick barracks on East 151st Street. Then they slowly drifted away. No one knew what happened to them.

Brenda didn't know all this history. She was only aware that the Emergency Assistance Unit had rejected her request for shelter three times in one month. She quickly surmised that the caseworkers at EAU were operating under some kind of unwritten quota. Of the dozens of families waiting for help, no more than two were ever granted eligibility per night as far as she could see. Never three. Never five. "They found every reason under the sun to say no," she said. She didn't have a good feeling about her prospects.

The irony here was that "homeless" was not something the thirty-seven-year-old mother relished on her résumé. Brenda had lived in her own apartment for ten years, holding down a revolving array of low-paying food-service, bartending, and cashier jobs. She made sure her daughter went to some of the best public schools in the Bronx, often springing for thirty-dollar round-trip cab fare when she couldn't escort her herself. Brenda had taken dental hygienist classes from a vocational school, hoping to end her on-again, off-again relationship with welfare. But she said her trouble started when her live-in boyfriend, Ty's father, smacked her around once too often. When she discovered he had a violent past, criminally violent, she got scared.

Roosevelt "Teddy" Jenkins cut a southern gentleman's pose the day Brenda met him at the deli counter in the Bronx. Brenda was dancing at a topless bar called Goat's in Hunts Point back then in the winter of 1992. She was bringing home fistfuls of off-the-books cash, sometimes four hundred dollars on up a week, enough to redecorate her apartment on a whim and keep her child in decent clothes. But Brenda was worn out from the night hours and the seedy scene. One morning she traveled up to Caldor's in Westchester to put in her application. On the way home she stopped at the grocery at Tremont Avenue and 176th Street for some cold cuts. The place was full of construction men on their lunch break waiting for their sandwiches.

A big fellow in a one-piece workman's uniform approached Brenda, offered his enormous hand, and said, "My name is Roosevelt Jenkins. How do you do?" Brenda noticed his pleasant smile and those courteous manners. So she shook his hand. "Please, I'll be waiting for you to call me," she remembered him saying. He handed her his phone number. The rest of his buddies laughed and teased him, calling him an old man.

Brenda went home, with no intention of calling him. She was not looking for a relationship. And besides, he was "high yellow," a pale black man. "I never went for no light-skinned men," Brenda said. "They have an attitude about dark women. Our hair isn't fine enough. They treat us with no respect."

Weeks went by, and the two ran into each other again at the White Castle hamburger spot. Brenda told him she really wasn't interested in going to a movie with him. Still, she gave him her phone number. He seemed very dignified to her, very polite. About a month later Roosevelt showed up at her door for their first date, dressed in a white linen blazer, summer shoes, a pink shirt, and a straw hat. "It was the dead of winter!" Brenda said, offended by his lack of fashion sense. "I don't like a man who doesn't know how to dress for the season."

At dinner the big man from Georgia charmed Brenda with his easy conversation. He talked about the things he wanted to do. He complimented her looks, calling her a "beautiful black lady." After a few months he moved into her Valentine Avenue apartment. At first he brought money and companionship into her life. Then drugs came on the scene. Beatings followed. An accidental pregnancy. Brenda wasn't sure she should have another child. Roosevelt wanted to marry her. He was happy about the baby. She knew he was trouble. Still, she tumbled forward, into motherhood, again.

"I let down my guard with Teddy. I had a wall up before him to protect myself," Brenda said. "Nobody could believe I could fall for someone who treated me like that."

One afternoon, when Ty was around six months old, Teddy came home early from work. She was preparing something in the kitchen, holding the baby in her arms. Teddy had an angry look. Something was off. "I asked him why he was home so early," Brenda told her attorney later. "He said he was checking to see if I had any men in the house." She snapped back, telling him he was crazy. Teddy then grabbed a butcher knife and pinned Brenda and Ty against the wall, threatening to kill her and then himself. "He said my life didn't matter. His either."

Teddy tried to argue that Brenda was the source of his trouble, polluting his life with drugs and her bar dancing. But he eventually backed away, leaving Brenda on edge waiting for the next eruption. One late evening, when Ty was one year old, Teddy charged home in a rage. Brenda could tell he'd been drinking. "This man had the devil in him, that's for sure. He'd been angry all week, calling me bitch, this and that. He was breaking me down. He was supposed to be my last stop," Brenda said.

Teddy demanded money. Brenda refused. He shoved her into the bedroom and started slapping her. Loreal could hear the lunacy from her corner in the kitchen. She never liked this guy, Teddy Jenkins. Her mother had called the cops dozens of times before when he started up with his temper. She couldn't believe her mother agreed to bear his child. It was obvious from the beginning that he was off the hook. The fifteen-year-old grabbed the phone to call the police. Then she stabbed his leg with something sharp—a pencil, Teddy would claim in family court. He hit the teen in the face on his way out the door, in self-defense he argued, fracturing her jaw. "He's a big old man," Brenda said. "Looks like Mr. Clean, with no hair on his head, and no eyebrows. Six foot two, punching my baby."

Police arrested Teddy. Brenda threw him out of the apartment. She pressed charges but later dropped them in order to move on with her life. He took a room nearby in the neighborhood. That's when Brenda found some of his court papers stowed in the back of her closet. To her horror, she read that her son's father had served twenty-two years in a Georgia prison for raping and beating a seventy-three-year-old woman and dragging her into her own yard to die. He'd left his hat under the woman's bed. Vicious and sloppy. He was on parole for life. And he still had a key to Brenda's apartment. She decided she had to flee.

There would be no nostalgic good-byes to this apartment in a central Bronx neighborhood. Brenda and her daughter had locked horns with some of the other tenants in their building over the years. One woman had them arrested twice for allegedly stealing jewelry. After three years of court dates, Brenda finally got the charges dropped. "The kids in the neighborhood were harassing my daughter," Brenda said. "Mothers were having their kids fight. Loreal calls it ghetto behavior, trifling. People who think they own the block. They were thinking we think we're better than them, or something. I don't know. I was afraid for my life, and for my daughter's, in that neighborhood."

Travails converged. Brenda's apartment changed managers, and her rent suddenly doubled. The landlord was making moves to evict her, bringing up a string of tenant complaints—from the same people she said were harassing her. She could have hired a lawyer to plead her case and keep the apartment, but it was no longer worth the fight. So on September 11, she called up her brother on Long Island and asked if she could stay with him for a while.

Loreal was seventeen by now, and she refused to live with her Uncle Walter more than a month. As far as she was concerned, he was a busybody who tried to run her life and put down her mother's. Nor did she relish the idea of transferring from Truman High in her senior year to some unfamiliar school on the island. So she moved to Long Island City, Queens, into a two-bedroom apartment with six people, her mother's friends. Brenda gave that family fifty dollars a week. The whole situation burned at Loreal. She started riding the subways everywhere, after school, sometimes during school. At night she accompanied one of her new roommates on her nightly rounds. The sixteen-year-old girl sold drugs on the streets of Long Island City for pocket money. The new welfare laws prohibited anyone under eighteen, even a mother, from getting her own welfare check until she had graduated from high school. The girl needed cash. Loreal didn't sell. She just watched. Brenda didn't know.

It tormented Brenda that her family had split apart. She wasn't one to dwell on the negative, but she blamed herself—her temper, her unfortunate choices in men—for messing up her daughter's life. So when Walter asked Brenda and Ty to leave before Christmas, it was a mixed blessing. He rarely hid his disapproval about his sister's predicament, calling her "nothing but a welfare this-and-that" whenever he could. When the two of them got into a sibling fight right before Christmas, Brenda packed to leave. Walter was sick of her mooching off of him. She was sick of his harangues. At least now she could have both of her children with her.

Brenda's next option was to place her family's future in the hands of the city. Her first stop was the Emergency Assistance Unit in the Bronx, the only homeless intake center left operating in all five boroughs. Security guards used a fluoroscope to inspect Loreal, then Ty, then Brenda, then their bags. He told them to discard their bottles of water and sandwiches. No food was allowed. Ty cried. He was hungry. The food at the EAU looked spoiled, but there was no choice. Ty ate it, and was sick all night. Brenda found some blankets in a closet and laid claim to a few square feet of linoleum space in the crowded "triage" room. The benches were mostly occupied. Men were fighting with their wives. Women were arguing with each other. Babies were hollering. Fluorescent lights buzzed overhead, all night. One little boy in a school uniform draped himself over his sister's stroller to sleep, his homework cluttered on the floor. A cluster of seven- and eight-year-old sisters hummed a homemade tune in the corner:

> *Hey, hey,*
> *You, you*
> *Time to go back to the EAU.*

Another family from Brooklyn arranged themselves in the same corner. Alice and Lorenzo Jones sat up all night on the iron benches talking to Brenda. Their sixteen-year-old son, J. J., and Loreal angled onto blankets on the floor. Alice kept repeating, part mantra, part joke: "The Lord takes care of babies and fools, so we must be fools." The intercom crackled, calling out names. J. J.—a tough homeboy in his East New York neighborhood—broke down in tears. Loreal steeled herself in her somber way, telling her mother under her breath that she would rather sleep on the street than spend another night in this place. "You have to be strong for me," Brenda told her.

Hours later Brenda's name was called. She sat down across from the Eligibility Assistance Unit worker and slowly took out her papers, one by one, laying them on his desk. His nametag identified him as a fraud investigator. This was a new job title at the Department of Homeless Services, the fastest-growing job on the city's municipal payroll. At a time when City Hall was slashing municipal jobs in record numbers, Giuliani had requested $1 million for staff increases at EAU for the previous year, and $1.5 million for the next year's budget.

The investigator told Brenda she should go back to her brother's house on Long Island. Brenda told him that wasn't going to happen. Four hours later Brenda was told she should gather her family and head for Auburn Assessment Center, while the city looked into her story.

Brenda and her children were accustomed to this city shuffle. They bounced from home to home, traveling the subways, filling out the endless papers that weighed down their suitcases. Auburn was just the latest setback. The patches of grass in this once graceful, cloister-like structure were barricaded from public use by a chain-link fence. There were bars on its arched windows, a new and ugly twist. Brenda had a way of holding her lean body erect, fixing her hair in a simple French knot, applying tasteful makeup—to distinguish herself from the crack addicts around her. She had a "Hello, how ya doin'?"

smile for anyone who cared. She wouldn't much notice those who didn't. Appearance was important to her, her armor against a disrespectful bureaucratic world intent on dragging her down. Sometimes the strategy worked against her—especially with black female case-workers who thought Brenda, one of their own, was acting too high and mighty. She knew the type. She didn't much like them. ("Black women are a trip," was one of her favorite sayings. Her negative leanings would come to haunt her.) Brenda tried to funnel her anger into a furious diary. It wouldn't be easy to keep her dignity intact in this place.

The cavernous, unpainted room assigned to them smelled of mold. Two windows were ajar, welcoming the frigid February air. She locked the door, wrestled the swollen windows shut, and started to work. Brenda had a thing about cleanliness. She figured it came from her third foster home, where the mother was a stickler for neatness. There had been some ugly scenes in that Long Island home, scenes that scarred both her and her brother Walter for life, but she considered the woman's orderly values a blessing. Brenda found some Lysol and started sterilizing the floors, the cot mattress, the walls. "I help you," Ty said, an eager volunteer in his mom's cleaning army. The two pulled some laundered sheets out of the suitcase. Brenda tacked one over the bare window. She turned around to find a male security guard in her room, staring at her from behind. "Turn around and get your sorry ass right out of here!" Brenda bellowed.

After a fitful night, Brenda roused herself at 5:30 A.M. to scour out the shower stall. Then she called Loreal and stood guard while her daughter rinsed, perched on stacks of towels to avoid contact with the floor. There were no shower curtains. Roaches lost their footing on the damp ceiling and fell onto the bathers below. The oversized, institutional bathroom was used by both men and women. It was impossible to be too paranoid under these conditions. Weeks later she learned that a fifteen-year-old son of an Auburn resident was accused of raping and sodomizing a seven-year-old he was supposed to be baby-sitting. This time there was a less disturbing episode. An old man down the hall from Brenda hurled a TV, a shopping cart, and metal cabinets out of his room. Emergency Medical Service workers dragged him off in a straitjacket.

For nine days Brenda waited for a letter to be shoved under her door from EAU summoning her back to the Bronx to announce its decision. By the time the letter arrived, she was warned, she might have only a few hours to pack her bags and make it to the Bronx on time. If she failed to show at the appointed minute, the city could use her non-appearance as an excuse to "log her out," as Brenda called it, the inevitable last step after "wearing me down." There was no time to look for a job, or take Ty to the doctor for his ringworm infection. Nor was there time, even, to apply for welfare benefits. Instead, Brenda spent the time with Lonny and Alice organizing the residents, while prodding the Auburn workers to clean the bathrooms, improve the cafeteria fare, and inform residents of the rules. J. J. came down with pneumonia. Alice locked herself in her room all day and cried. Brenda felt imprisoned.

Finally, the letter arrived. She hauled her bags on two subways back to the EAU. Rejection. Twice more she made the trek. Twice more, rejected. After the third time, she abandoned her vigil and sought out legal help. She had heard tales of people being sent back twelve, fifteen, even eighteen times. That was not going to happen to her, she vowed. In her journey around the system, Brenda had stumbled across a fledgling welfare organizing group called Community Voices Heard. The group, staffed by welfare recipients, was dedicated to finding ways to fight the new regulations. Two local television news reports had carried sound bites from Brenda talking about her experiences with the new welfare rules—her three-month wait to get a welfare check, the diversionary tactics in the homeless way station. CVH hooked her up with a pro bono attorney to help plead her case.

Emboldened by her newfound advocates, Brenda made a third trip back to the EAU. She left her bags at Auburn this time, determined not to be sent packing again. The atmosphere in the dead-end outpost had decidedly changed in her favor. "Suddenly it's 'Miss Fields' this, and 'Miss Fields' that. 'Wait here, Miss Fields.' 'Don't worry, Miss Fields.' 'How do you feel, Miss Fields?' " Brenda found this new respect disingenuous, the result of her recent fame as a low-level celebrity with real legal representation.

"How do you think I feel?" Brenda hollered back. "My daughter is trying to go to high school in the Bronx. She is trying to get an education. We're living with schizos and outpatients. It's not safe. I can't begin to look for a job, or even get on welfare, when you have me waiting all day for letters and running all over the city to make appointments."

After thirty days in Auburn Assessment Center, the authorities decided to help the Fields family find temporary shelter. EAU had been wrong to throw her case out. Brenda, Ty, and Loreal were officially homeless, something they'd known all along.

Brenda lowered her canvas bags again onto the sidewalk on East 138th Street in the Bronx, back in her home borough, resting one more time before ascending the stairs to the Jackson Avenue Family Shelter. It was March 13, 1997, nearly two months after her first request for help. Across the street, churchgoers were filtering in and out of the Iglesias de Dios Pentecostal storefront. Gospel tune fragments breezed out into the street with each swinging of the door. This was certainly a Pyrrhic victory—entry into a homeless shelter—but Brenda decided to feel good about it anyway. She had no money, no food. She had tried to apply for welfare benefits, but had so far been unsuccessful. Still, Jackson looked like a suburban housing project to Brenda and her kids. In fact, the red brick building was only ten years old, an innovation in homeless housing. Children played in a padded outdoor area. Beds of impatiens dotted the front yard. The yellow and blue tiles in the hallway looked cheery, freshly scrubbed. The building housed a medical clinic, a job center, social workers, and a big community room for meetings and social events. Ty poked his head into the early childhood center, entranced with the crayons and the trains, his favorite obsession.

If she hadn't fought so hard to get here, perhaps she would have noticed its more ragged edges, its infantilizing touches. Brenda didn't really see the midnight curfew warning for adults on the front door, or the younger woman crying as security guards forced her to dump out the contents of her purse onto the front desk. After a few days Brenda would clamor for air in her tiny room, just big enough for two single beds—one for Brenda, one for Ty and his older sister to share. Someone told her a previous tenant had operated a hooker business through their first-floor window. Ty would spend hours leaping from one bed to the next. There was no floor space to play. Residents banged on doors all day and night.

Disorder seemed inevitable. Brenda found herself embroiled in sometimes violent neighboring dramas. A wife threatened to stab another resident for carrying on a blatant affair with her husband. A mother fought trumped-up child abuse charges. A building janitor was running a loan shark business that ensnared the child care staff. But on this first day Ty broke loose from her hand and bounded down the hall, trying to find a way outside to climb on the brightly colored plastic playset.

Loreal got busy unpacking, happy at least to be closer to school. Graduation was just three months away. She'd missed a good chunk of coursework with all this nomadic living. She wanted to walk across the school stage with her class to get her diploma. If she made it to all her exams and finished all her credits, her mother would never stop bragging about it. Loreal might not be able to buy a prom dress or pay for senior trips. She might not be able to attend her high school dances with the kids' curfew at 10:00 P.M. At least this tiny room had to be better than the decrepit conditions in Brooklyn.

A tall, gentle man greeted Brenda and her kids on that first day. He struck Brenda as almost too kindly, too pensive, to preside over this South Bronx cauldron of domestic stress. "Welcome to Jackson," he said, haltingly. There was no edge. None of the disrespect to which she'd become accustomed. "I am the director of social services here. Please, make yourself comfortable."

Hank Orenstein, with his beatnik beard, Moroccan cross, and thoughtful sidelong glances, provided an ephemeral pause in this world of tension and tempers. Instead of slumping into his paperwork, Hank would jump to his feet when a resident entered the room, eager to help. He greeted everyone warmly, giving his full attention. As Brenda liked to put it, "Hank's cool. He can calm the meanest bull." When a fight broke out between two women over a baby's missing jewelry, someone shoved Hank against the wall when he tried to intervene. Unfazed by the head banging, Hank simply said, "Let's everyone come into my office and talk." There the brawl melted into a whimper.

Hank was brought in to head Jackson when the city contracted out its leadership to Citizens' Advice Bureau, one of thirty-six nonprofit social service settlement houses in the city. The first thing Hank did was put screens on the windows, paint the walls, exterminate the pests, and wire every room for phones. Then he set up an unprecedented battery of human services—a medical clinic, an after-school program, parent classes, domestic violence counseling. Jackson was still a place taut with tension, breeding what Hank called "toxic social relationships," but it became much better than it had been. About 80 percent of its children attended school regularly. And nearly 10 percent of the ninety-five families moved into permanent apartments every month.

Hank was to become an important person in Brenda's life. She appreciated how he smoothed over her first days. He helped her stock her room with food from a nearby pantry. He helped direct her to the right city welfare center so she could begin to receive public assistance—a small cash allowance, food stamps, and Medicaid health insurance. He marveled at how she coped with the stresses of Jackson, greeting other residents as if she ran the place. She helped neighbors when they ran into bureaucratic walls, dressed down staff members when they got fresh. He recognized her latent leadership talents. He asked Brenda to speak to his Saturday social work class at Fordham University. When she wrote a résumé, his name would be the only reference.

The two made an odd pairing. Brenda's tough edges contrasted with Hank's malleability. Brenda bellowed, while Hank soothed. She couldn't help but find some fault with his accommodating style. She thought he should rule Jackson with a firmer hand, clamping down on power-tripping security guards who commonly charged unpopular residents with rules violations, thus jeopardizing their welfare checks.

Hank's heart wasn't in such sheriffing. He didn't become a social worker to police the homeless and help knock them off the dicey welfare rolls. More and more, that's what his job was becoming. Trained twenty years earlier at the relatively pacific University of Iowa, Hank had worked in social services both in Iowa and in New York City for much of the past two decades.

Most residents didn't realize Hank was an aspiring artist during his off-hours, specializing in landscape microphotographs—playful mergings of paint and organic matter, magnified through a microscope. He called the technically intricate designs "visions beyond the eye." There was a disturbing quiet to his nature landscapes—a dark, solitary beauty.

"I'm part of the problem here," Hank said one day in his cluttered office, overrun with Crayola markers, subway maps, research documents, and photos of his six-year-old son. Why were these social services available only in a homeless shelter, he wondered? Why did the city find it reasonable to spend three thousand dollars per family a month to keep them in a shelter and spend next to nothing to provide permanent housing? "Shelters like this shouldn't exist," Hank insisted.

Instead, Hank said the city should have real housing plans. Since Giuliani had taken office, the city had cut the number of apartments it supplied for homeless families by 92 percent. Apartments for people with AIDS were cut in half, even though the AIDS rate for city residents was growing. The city wasn't connecting social services with churches and communities, Hank felt. It should have created jobs instead of forcing welfare recipients to sweep the streets for their checks. It should have built more low-income housing instead of more shelters, with more temporary beds.

"Housing costs have gone up 250 percent in the last decade, and yet we are giving a family of three $286 a month for housing allowance. The same family needs a wage earner making $13.50 an hour to pay market rents. That's not happening for these families either. It's irrational," Orenstein said. "I don't see any vision or hope with the current welfare reform. It devalues the parent who wants to stay home and raise her children."

He had dedicated himself to improving the lives of the poor. But the more he contemplated the larger picture, the more he felt compelled to move on. Hank began planning his next career move, organizing tour groups to the outer boroughs, or directing a policy program. He exhibited his art in SoHo galleries, he taught social work at Fordham University. He attended community conferences on welfare and housing. He stayed active in the cerebral fight, while he waited for his escape date.

One day in April, Hank made a decision that unwittingly nudged Brenda's life a notch forward. Her welfare case had finally opened two months earlier. Brenda was putting in her workfare hours at the shelter's day care center, cleaning up the snack area, designing new bulletin board displays, tying shoelaces. While other residents were busy picking up trash and cleaning subway stations in exchange for their benefits, she had managed to set up her Work Experience Program duties at Ty's day care center. Hank glided into Brenda's view. "Brenda," he said haltingly, "I wonder if you would be interested in speaking at a conference coming up. It's on the violence of poverty. I thought you could share your story." Brenda didn't have to think about it long. Tyjahwon's father had just paid an unannounced visit to Jackson a few days before, trying to convince the security guards to let him visit his son. Brenda was trying to take out an order of protection notice against him again, but she didn't know his address. So, yes, she would talk at the conference. The violence of poverty was something she was familiar with.

A few weeks later, on May 19, 1997, Brenda found herself seated on a stage at Hunter College Auditorium in Manhattan, looking out over a large audience of students and social workers. She had found her voice in the hallways of Jackson, teaching residents about the new welfare time limits, helping them fill out all the forms. But public speaking in a room full of social workers was something else. A kindly, academic type sitting next to her on the panel said, "Are you nervous? Just look at me when you talk, and you'll be fine."

Brenda started off with her handwritten speech, haltingly. "Good morning. My name is Brenda Lee Fields. I'm the proud mother of two children. I've lived in the Bronx for twenty-five years." She looked up. The woman who had given her advice nodded her along. Her voice grew in confidence. "Back in August '96, some financial and personal problems became a threat to my children and myself. I was not working. I was receiving public assistance. I felt myself sinking, living from check to check, falling deeper into debt, robbing Peter to pay Paul, so to speak. I had no luck finding a job paying decent wages to support my family. I found the welfare system to be degrading and very stressful. If it weren't for my faith in God and the love for my family I would not have made it this far. . . ."

It was a turning point for Brenda. When her speech was over, the woman, Barbara Petro-Budacovich, asked Brenda to meet with her. She was opening a unique housing complex on East 168th Street in the South Bronx for low-income families, and she was looking for residents like Brenda who would be an asset in the building. It was going to be a full-service apartment building, with day care and job training services run by the Women's Housing and Economic Development Corporation.

Brenda couldn't believe what she was hearing. After more than a dozen trips to the South Bronx welfare center, papers in tow, she finally had begun to receive a welfare check in March— $145 cash every two weeks, plus $265 in federal food stamps per month. It was standard fare and actually would be cut, without warning or reason, in the months to come. It was barely enough for the three Fields to live on in the shelter. She didn't know how she was going to pay market-rate rent. In the ensuing months Barbara helped smooth out Brenda's paperwork, getting her a Section 8 federal subsidy to pay the full $465 a month for this new apartment.

On September 11, 1997, Brenda and her children packed up their canvas bags one last time. It was exactly one year to the day since they began their bumpy odyssey. Ty's "Cat in the Hat" was now too worn to travel the few blocks north. Brenda still had the Avon bag, filled with her proofs of existence. Loreal added her high school diploma and a set of graduation photographs to her small collection. The three Fieldses lugged their possessions up yet another set of stairs. These concrete steps, at East 168th Street and Walton Avenue, were by far the most impressive. Constructed in the twenties, when entrances were designed for grandeur rather than ease of entry, the wide concrete stairs signaled a return to some dignity.

Brenda walked to the marble counter and signed in to the security guard book. Like Auburn, this building was once an abandoned hospital with a tale to tell. It had been converted, however, with care. They walked across the renovated terrazzo floor in what was once the hospital visitors' center, past its Florentine windows, and into a freshly scrubbed elevator. Brenda poked her head around the elevator bank corner to see which hallway was hers. The linoleum floors nearly blinded her with reflections from the recessed lights. Their apartment had two bedrooms, a living room, a walk-in kitchen, and ten-foot ceilings. It seemed palatial after their first-floor shelter dormitory. The walls were a freshly painted beige. The kitchen cabinets were oak— never before used. The windows were tall and arced, welcoming the daylight and the view from the North Bronx. Ty looked out and saw the No. 4 elevated subway train snaking by, almost toylike. "It's the Bidicudous Train," he said, his personal favorite. Yankee Stadium was just a fifteen-minute walk west. Brenda tacked up a portrait of Jesus in the hallway, got her couch and TV out of storage, and set up house.

There was little time to celebrate. In the next few weeks Brenda was visited by a whirl of inspectors. A city fraud investigator from the Eligibility Verification Review office made certain Brenda wasn't harboring any able-bodied freeloaders at city expense. A federal inspector checked to see that the apartment was up to code. A shelter caseworker inspected Brenda's housekeeping skills by lifting up the toilet seat with her pen, disapprovingly. Brenda chased the details, kept her temper in tow, and continued to feed the system's insatiable appetite for paper.

All the while she was keenly aware that her welfare clock was ticking. Nearly half a year was gone already. In four more, she would have to pay for everything herself or lose this small piece of urban heaven forever. The new welfare laws placed a five-year lifetime limit on benefits. There was no margin for error. The rules also stipulated that half of any state's recipients had to be working for the benefits by the rules' end. City welfare agents would be forcing her to work another WEP job any day now. Nothing bad could happen to her again. Then her buzzer rang once more. This time it was Teddy Jenkins, delivering a court order for Brenda. He was filing for custody of Ty.

2

Brenda

Black Ostrich

Nuns kissing, hugging, cooing, tickling, fussing, kissing her some more. These were Brenda's first memories. She couldn't have been much more than one year old, a baby cast adrift along with her older siblings, Walter and Louise, the eldest.

The year was 1961. Her mother had been sent to Creedmoor Psychiatric Hospital at the young age of twenty-one. The children's father had long ago hit the road, leaving his fragile wife before baby Brenda was born. No one knew who he was, nor why he left. Or if they did, they weren't saying. The children only knew him as a line on their birth certificates:

Father: Walter Fields
Race: Colored
Occupation: Laborer
Birthplace: Estel, South Carolina

Grandma was the emotional rock of the family. She cared for all three grandchildren as long as she could, all the while watching her own young daughter drift in and out of mental illness. Then the old woman's body just gave out. A clot cut off the blood flow in her brain.

The children were as good as orphans. The Children's Aid Society became their new caretaker, taking over the task of finding a foster home that would raise all three together. That was what the agency told Brenda forty years later when she wrote to inquire. Brenda never knew her mother, or, of course, her grandmother. She had no sense of how either woman smelled or walked or laughed. It was a vacuum in her memory. She spent very little time filling it with fantasies. Brenda thought maybe she was more like her grandmother, soldiering on in the here and now despite the stresses and the unreliable heart. She was forging her own truths, her own family ties.

One day a big red Pontiac with chrome detailing and flashy fins pulled up to their Catholic refuge. The woman behind the wheel told the Fields children they were leaving the nuns to join a new family. This first foster mother was a very gentle woman, young and disciplined. She dressed the children in their Easter finest for church. Brenda has a picture of the three of them, posing in starched pastels, grinning in their patent leather shoes. The woman's husband was kind to the children, but he had a violent streak. Once he whipped his wife with an extension cord while she was bathing Brenda in the kitchen sink. Brenda was probably around two years old at the time. Brenda's next memory was of the woman hanging curtains in the living room, next to photos of Martin Luther King Jr. and President John F. Kennedy. The phone rang. Brenda's foster mother gasped, and began to cry. Kennedy had been shot. She scooped up young Brenda and took her to a friend's house so the two women could cry together. Soon after, the foster mother gathered the three children in the living room and told them she loved them and she would always love them, but she had to give them up. Hers was not a good home for them now. She needed to get her marriage together.

So the Children's Aid Society trundled Louise, Walter, and Brenda off to their next home—a rambling, two-family house in South Ozone Park, Queens. This house, with its elegant bay windows, its cellar stocked with wines, its spacious bedrooms would be forever burnished in Brenda's imagination as an enchanted Eden, her ultimate salvation. Mr. and Mrs. Jones (Brenda couldn't remember their first names, if she ever knew them at all) were a kindly couple in their late fifties. They had longed their whole lives for children to fill up the sumptuous space near the Van Wyck Expressway. Brenda remembered climbing the stairs to see her room for the first time. In it was a child-sized rocking chair. A stuffed lion and other soft, squishy animals sat on the bed. She'd never seen such attention paid to a child's sense of comfort. In the yard a canopy swing hung under arching grapevines. Tomatoes and cucumbers climbed the wall of the two-car garage. A pear tree grew in one corner of the yard, a fig tree in the other.

Christmases were lush with gifts and decorations. The tree seemed to fill up the entire living room, right to the ceiling. Under it one year was a bicycle, just for Brenda. The home had all the makings of a fantasyland for these motherless three. "The old man just loved us to death," Brenda said. "When the neighbor kids would come teasing, calling us foster kids this and that, Mr. Jones would come charging out of the house, yelling that he would knock their blocks off if they ever bothered us again." He would always talk about sending Brenda, Walter, and Louise to college someday. Then tragedy struck. Mrs. Jones was diagnosed with cancer. It spread rapidly through her body. She told the children not to worry as they gathered around her deathbed. Her husband would not let them go. She told the children a relative would come down from upstate to help Mr. Jones care for them. But in the end the agency would not allow a single parent to care for foster children. In the late sixties, it was against the rules. "Mr. Jones broke down when they came to get us," Brenda said. "He was a big man, and he broke down and cried."

Brenda was seven years old when she first met Lillian and Roy Eskridge, her new foster parents in central Long Island. The first thing Lillian did was to ask the children and their foster care agent to remove their shoes. She was a fastidious housekeeper. Then she led them to see their rooms in the large two-family home. Walter was to sleep in one room. The girls would share another.

At first Brenda was eager to help with the seemingly endless chores. At age seven, she was ironing sheets, handkerchiefs, and pillowcases. At age eight, she was taking apart venetian blinds, soaking them in soapy water, and restringing them in the backyard. By nine, Brenda was caring for Lillian's infirm sister, Ida. Before she went off to school in the morning, the fourth-grader would bathe the old woman, fix her hair, feed her, and empty her bedpan. Walter mowed the lawn, raked the leaves, and fixed the plumbing. Louise was set to work making clothes for everyone. Whenever Lillian had a party, the children would serve the food and clean up afterward. Whenever Roy brought home bags of clothing discards from his elevator job at Bag Dag Clothing, Lillian would hold a yard sale and ask the children to sell them. Foster care sent pocket change every month, so the kids could get their hair cut and buy a soda or two. Brenda does not remember seeing any money, even when they became teenagers. Perhaps most cruelly, Lillian kept a lock on the refrigerator, a lock on the freezer, and a lock on the phone, to keep them in check. "It's funny the reasons why some people want foster children," Brenda said years later. "In her case, she was just real greedy."

Brenda attended junior high in Center Ridge, a school with fewer than thirty black students at the time. Being the only black child among whites took some adjusting. The Fields children were accustomed to all-black schools in the city. Now the only time Brenda saw other black children was during lunch period. She was the target of a few racist remarks, but mostly silly taunts. "Country Girl" she could stand. But "Black Ostrich" made her mad every time. She got into several tussles on the bus when kids would crank their heads in and out like overgrown chickens, mocking Brenda's long neck and her long legs. Two more incidents caused her to avoid taking the bus altogether. Once a friend of Brenda's brought her uncle's real Ku Klux Klan costume on the bus as a novelty item. The girl had no idea the black kids would take offense. Another time the bus pulled up to the school's front door. A hush fell over the morning chatter. Someone had scrawled "Nigger Get Out" next to the entrance. No one moved. Brenda felt everyone staring. Finally, the principal came out to greet the bus and to escort the black children one by one into the building. After that Brenda walked to school. This way she never felt outnumbered. But more to the point, she used the walking time to smoke cigarettes. Her brother Walter was a champion track star. She hoped his athletic prestige would form a magical shield around her. It never worked out that way.

The children saw their biological mother once during their years in foster care. The Children's Aid Society arranged a meeting at its Manhattan office with the children, their mother, Eleanor Odom Brown, and her brother, Earl. Brenda was about ten years old. She remembered her mother as a shadow in a sweater, with blank, staring eyes. Eleanor kissed the children as if in a trance. Then she sat down and stared at the wall, her head wrapped in a scarf tied under her chin. She seemed drugged, not altogether there. Uncle Earl introduced her. "This is your mother. She loves you very much," he bellowed, like a Pentecostal preacher. "Uncle Earl promised she would come get us one day," Brenda said. "And he also promised to send a box of clothes real soon. He took down all our measurements. Like fools, we ran to the postman every day for a long time." The box never came. The children took Earl for a charlatan. They never saw him or their mother again.

Meanwhile, things in the Eskridge house deteriorated. Roy began groping Brenda sexually on the weekends when he came home from work. She threatened to tell his wife and the world if he didn't stop. Brenda was determined to get this situation fixed. She tried to figure out how to alert the agency that the Eskridges treated them more like servants than children. The trick was finding a way to do it without placing themselves in the path of Lillian's wrath. But the Children's Aid Society began catching on to Lillian's schemes on its own. First, the caseworkers noticed the children were afraid to talk when Lillian was in the room. In fact, Lillian would threaten them whenever the caseworkers left. Then they noticed her expenses didn't add up. Lillian would charge the Children's Aid Society for train fares when she drove the children into the city for interviews. The caseworker would ask the children about their train ride. The children would say, "What train? We drove in a car." Or Lillian would submit expenses for bicycles. Again, the caseworkers would inquire about the bikes. The children would shrug their shoulders.

Lillian always said no one else wanted the children except her. She was the only one who would take all three. She always promised to adopt them one day. They believed her. They were terrified of being split up, of leaving the only home they knew. And they figured that after three foster homes, who else would want them? But finally, Lillian pushed too hard. Brenda confronted her. They fought. Soon after, the Children's Aid Society moved fifteen-year-old Brenda to her fourth foster home, this time in the Bronx.

Louise eventually graduated from high school and was off on her own. Walter remained in Lillian's home. The foster mom convinced him to stay and take care of her. In exchange, she told Walter, he would inherit her house and all her property. He did stay, Brenda said, working hard on her houses, nursing her during her own fatal bout with uterine cancer. One day Walter found a drawer full of letters addressed to him from colleges offering him track and field scholarships. Lillian had hidden them fearing she'd lose her free help if he went off to college. Walter left the house in bitter haste to join the marines. When Lillian died, she left Walter nothing. "That's why he's so coldhearted now," said Brenda. "It's the way he was treated."

Brenda's Bronx foster home was supposed to be an emergency placement, yet it lasted for nearly four years. She never felt wanted in the Martin house on 213th Street. She was always the third girl out. Mrs. Martin already had one foster daughter, Dorothy. Her own daughter lived in the apartment with her small children. Brenda's Cinderella status was most obvious around the holidays. Dorothy would get a camera under the tree. Brenda would get an umbrella. To make matters worse, the home was filthy. Food was left out on the table uncovered. No one dusted or cleaned. If there was one thing Brenda had gotten used to in Lillian's house it was a spotless environment. Brenda left that last apartment when she graduated from Fashion Institute High School. The city stopped making payments to foster parents when their charges turned eighteen. Mrs. Martin invited Dorothy to stay. Brenda had to go. She'd fallen in love with a gangly musician in the neighborhood, a gentle percussionist with his own band and two West Indian parents who lived in a nice house around the corner. Brenda moved in with him. "He was the only one who really cared for me," she said. At age nineteen, Brenda became pregnant with Loreal.

In many ways, the sad day when Brenda, Walter, and Louise left Mr. Jones crying on his lawn as he waved good-bye to them sealed each of their fates. Brenda often wondered how their lives would have turned out if they hadn't been wrenched from utopia in Queens. She would have gone to nursing school. Walter surely would have become a college track star somewhere. Louise might have studied business. Most likely none of them would have spent a moment on a welfare line. Brenda slowly lost touch with her siblings over the years. They stayed on Long Island. She never moved from the Bronx.

Brenda always felt different from her brother and sister somehow. She was always bigger-boned and louder-mouthed. The lore among the three children was that Louise had a different father. Her last name was Brown, like their mother. Brenda and Walter were Fieldses. But after forty years Brenda discovered that it was she who was spawned by a different gene pool. The Children's Aid Society told Brenda that Linwood Rhodes was her dad. He was a part-time trucker and occasionally helped the family out with money. Perhaps the specter of Linwood in the apartment door had caused Walter Fields to hightail and run all those years ago. She would never know. Whatever the case, Brenda kept the Fields name and attached her imagination to the only one who had shown some gumption in the family—her grandmother, a soul mate whose face she couldn't even conjure. Grandma was the one who fought to hold the family together. She must have known how to be a strong role model for her kids. She hung on until she died. With no other decent parent in Brenda's life to learn from, an apparition, without a name, would have to do.

3

Alina

Accidental American

Through her airsick wooze, Alina Zukina tried to sweep the blur of buildings whizzing past her car window into a first impression, a feeling for America. It was a shadowy autumn day in 1994. Alina and her family had just fled the former Soviet outpost of Moldova for New York City. The family had waited seven years for this day. They gave up their coveted government apartment with three balconies, one that had required more than a dozen years of slow rising on a waiting list. They sold all their books and furniture—everything they owned—to finance the exodus. "We work hard our whole lives, and we leave with nothing," said Alina's mother, whose forty hard years showed like tree lines around her eyes. "Only our dignity."

And now, traveling north in her uncle's car on the Van Wyck Expressway, the elfin nineteen-year-old felt only disappointment. Where were the glass and steel skyscrapers, the glad-handed crowds of New Yorkers darting about in furs and Armani suits? Where were the glamorous window displays, the gold-plated statues? Hers was a common Eastern European fantasy vision of the United States, conjured by post-glasnost Soviet television, which broadcast almost exclusively from Rockefeller Center. America's glitter and wealth were supposed to relieve her family instantly from the sting of deprivation. No more stoking coal for heat, no peddling used clothes on the streets, no storing potatoes on the balcony for the winter, no rationing electricity or hot water. Russian-style anti-Semitism would no longer rein in their futures. Alina's dad would find work as a machinist, maybe in one of those impossibly tall office buildings. Her mom would supervise a bevy of kitchens. Her nine-year-old brother would get an American education.

Alina would finally become a doctor.

That was the dream anyway. Pushing her cherubic face against the car window, Alina watched as the squatty buildings, tangled electrical wires, and low-lying red brick row houses sped past. This was Jamaica, Queens, the residential neighborhood surrounding JFK International Airport. She squinted, hoping to pierce the haze hovering over the rows of working-class and low-income apartment buildings to catch some inspiration from the towers of American affluence. There was none.

Alina tried to tamp down her persistent anxiety. Maybe her family had made a big mistake. It was on her mind for the dreadful thirteen-hour plane ride—her first, and, she hoped, her last. Her brother, Moshe, dozed beside her in the car, his light hair still sticky from the juice she accidentally spilled on him before they boarded for America. Just days earlier the Zukinas had sat for the last time on the floor in their empty apartment in Kishinev, Moldova's modest capital, to discuss their nervous options. "Maybe if it doesn't work out, we can come back," Alina's mother said. But that prospect was unrealistic. The family understood there would be nothing in their homeland worth returning to. Moldova had been in financial disarray since 1991, when a nationalist movement declared the remote slice of land wedged between Ukraine and Romania independent from the Soviet Union and established a free market economy overnight.

Alina was a reluctant refugee at best, uprooted by economic uncertainty in the wake of the Soviet Union's collapse. She had never imagined life beyond Kishinev's potholed boulevards and hurly-burly markets. She had spent her nineteen years roller-skating on its streets, strolling in Pushkin Park with her friends, lecturing her mother on the value of conserving water, studying hard to make straight As from her strict Soviet teachers. Her mom would smile and tease, calling her "my good little Communist." She wanted her brainy daughter to become a good homemaker, a good wife. But Alina's only dream was to work as a doctor in this country that was almost Balkan, almost Russian, but most certainly Eastern European. She had no burning desire to embrace a culture outside the odd and volatile stew of Romanians, Russians, and Jews that made up the 4.4 million residents of Moldova. But when Moldovan nationalists swept the old Soviets out of power, her way of life was swept away too. The new economy created opportunity, but also uncertainty for all—for the occupying Russians, but most of all for the country's Jews.

Her mother was laid off soon after the revolution from Cafeteria No. 22, a supervisory job she'd held for fifteen years. The new government couldn't afford to pay all the salaries. Lyusya Abramovna Zukina, the only Jew, was the first to go. She scraped together a few *lei* during her years of unemployment by traveling to Romania to sell used clothes and other trinkets on the sidewalks. She and two housewife friends would share one bed on these vending jaunts, sleeping in their clothes. Boris, her husband, was the first to be fired from his agricultural factory position when it was partially shut down. He would make his way by train to Moscow for months at a time to pick up handyman jobs. Despite her stellar exams, Kishinev's Medical Institute rejected Alina's application, dashing her hopes of becoming a physician.

All three believed they suffered these setbacks because they were Jews. There was no hiding their identity, no matter how much Alina sometimes wished she could. Under "nationality" on their Soviet passports, on all documents, officials wrote "Jew" or "Jewess." Kishinev's Jews, many of whom were from Ukraine originally, were Russian speakers, not native Moldovan speakers from Moldova (an ethnic group akin to Romanians), so they were not considered an integral part of the new nationalist fervor. No matter what the new government said officially, everyone knew that jobs and university positions in the provisional country would first go to Moldovans, who spoke the newly official language.

Several incidents underscored the Jews' tentative place in the new Moldovan republic. Soon after the bloodless revolution, Kishinev's Holocaust memorial was doused with red paint and sixty tombstones were destroyed in the old Jewish cemetery. The heavy plaque marking the Jewish library across from Alina's home was torn down again and again. Graffiti demanded that Russian occupiers retreat: "Moldova is for Moldovans. As for Jews, you know where you should go." Once, a rumor spread throughout the capital that a pogrom was going to sweep through Kishinev's Jewish homes.

The conjunction of "pogrom" with "Kishinev" could not have been more chilling. On Easter Sunday 1903, mobs of Christian youths and laborers had rampaged through the Jewish sections of Kishinev, ostensibly to avenge the "blood libel" of a fourteen-year-old Christian boy whom the mobs believed was killed for Jewish Passover. For two days they battered Jewish shopkeepers, bakers, and glaziers with axes and crowbars, tore through Jewish houses, raping women, smashing their china and furniture, ripping apart their feather beds. All the while the police stood back. When the mayhem died down, one writer described an eerie scene of small white feathers settling over the carnage like a blanket of snowflakes. What looked to the world like a Russian-condoned killing spree was condemned by Leo Tolstoy,

Maxim Gorky, and the press in the United States and Britain. The real spark for the pogrom lay in financial decline and a growing Moldovan nationalist movement that sought to expel Jewish outsiders and the Russian imperialists from this Bessarabian village. Jews emigrated to America in droves.

The historical similarities spanning the century were too horrifying to contemplate. Alina's family bought enough food for several days and barricaded themselves inside their apartment, shoving furniture against the doors and windows, waiting in fear for the violence that never came. The nationalist zealotry would subside in later years as the entire country careened deeper and deeper into debt. But for now there was no choice. The Zukinas had to leave. The rest of their family was already in America.

Alina's uncle drove on, over the Triborough Bridge, up the Major Deegan Expressway, over the Cross Bronx Expressway, and up the Bronx River Parkway to Parkchester, home to newly arrived Bangladeshis, Dominicans, and five thousand Russians Jews, one of the smallest former Soviet enclaves in the city. The green and rolling parkland of the Bronx was once the desired destination for half a million Jews who had made enough money to escape the stifling railroad tenements on the Lower East Side. Now fewer than ninety thousand still made the Bronx their home, and one-third of them lived in poverty. Even so, Russian Jews were providing a mini-boom for the borough's collapsing synagogues. Alina's uncle had emigrated here with his family several years earlier, and he now made a decent living driving a limousine, while his wife worked as a home attendant caring for elderly shut-ins. The couple had fled Ukraine in fear in the late eighties after being targeted by a vocal group of anti-Semites, who were emboldened by poverty and the failures of perestroika. When a Sharogrod laborer became ill from food poisoning, villagers blamed Alina's aunt and uncle, who supervised the local meat packaging plant.

Sharogrod residents had marched in the streets demanding that the demon Jews "go back to Israel."

The Zukina family scrambled out of the car onto Pelham Parkway, a tree-lined Bronx boulevard of low brick apartment buildings and bustling shops. "Babushka! Babushka!" Moshe hollered when he saw his grandma, Basya Burd. He repeated his cry until it finally penetrated her hearing aid and she hobbled up to give her grandson a hug. Lyusya's mother had fled Ukraine in the early nineties, with a wave of more than half a million Russians, many of them elderly. Old-age pensions in Russia amounted to less than twenty dollars a month. Many seniors received nothing for months at a time because the government couldn't afford to cut the checks. It was emigrate or starve. In America, Basya could live on Supplemental Security Income for the disabled, with added help from local Jewish philanthropies.

The four newcomers to America possessed six canvas bags, some pots and pans, a set of china, a little cash, and a few words of English. They wondered where they would hang their rugs to beat, and where Moshe would play outside. They wondered how they were going to raise $675 every month in rent for the one-bedroom apartment upstairs from Grandma Burd. Would they find the salty *brinza* cheese, the sour *vishnya* cherries, the *kielbasa* sausages, and the *sala* pig lard they loved in the markets here? Would they learn the language and find jobs? The unknown filled Alina with some resolve, but mostly terror.

New York City has been both the yoke and the salvation for millions of Russian Jews since the nineteenth century. Thousands flocked into Manhattan's Lower East Side tenements at the turn of the century, fleeing Cossacks and pogroms precipitated by the collapse of czarist Russia. They worked in sweatshops for builders and for storeowners, helping to construct the city's infrastructure and clothe its citizens. They labored in industries that no longer dominated city commerce. Hundreds of thousands more fled Nazi concentration camps in the 1930s, becoming American professors, department store heads, pickle shop owners, scientists, drivers. A 1989 federal amendment lowered entry standards to America for former-Soviet-bloc Jews and evangelical Christians who could establish even flimsy proof of a "well-founded fear of persecution." As a result, Russian Jewish immigration surged when the Soviet Union collapsed. In the nineties, more than eighty thousand set down roots in the city. From 1990 to 1998, one in eight of all the newcomers to New York City were Russian Jews.

Russians, in the government's view, have almost always been "desirable" immigrants. Russian Jews' high education levels, pursuit of religious freedom, and arguably Caucasian race were important factors in their favor. By contrast, the year Alina's family gained refuge in America, most Haitians were turned away. Though they were fleeing a regime accused of assassinations and torture, only eight hundred out of the thirty thousand Haitians who applied in 1994 were granted asylum in the United States. That same year, 90 percent of Russian Jewish applicants were granted visas.

Russian Jewish visas were the envy of most other immigrants. The vast majority, along with Vietnamese, Cambodians, Thais, and Laotians, were labeled "refugees," the most coveted designation for new Americans. Refugees were allowed four months of federally subsidized support, including language and job training. They also had the option of turning to public assistance immediately thereafter. Other legal immigrants were told to wait for five years before applying for basic aid. Another factor in their favor was the network of organized and well-funded philanthropic groups that helped Russian Jews finance the exodus from their homeland and adjust to their new one. The groups included the Hebrew International Aid Society, the United Jewish Appeal Federation, the New York Association of New

Americans, Jewish community councils, and the Metropolitan Council on Jewish Poverty. Though more than one-third of Russia's Jewish refugees went on welfare after their initial three-month grace period—the highest percentage of any immigrant group—most then fell off the rolls in a rapid two years' time. The reasons, experts believed, were education, strong families, and reliable support groups.

If Alina and her hardworking parents were considered the cream of immigrant welfare recipients, that status was lost on them. Everything in America was harsher, more expensive, and less accessible than they had expected. Buying food, applying for benefits, just getting around, were all treacherous challenges.

Life was disorienting and disheartening to a foreigner in an adopted land, filled with people of unfamiliar colors and cultures, speaking a different language. The family felt exposed at first, like neon Teletubbies on the set of *Homicide.* Alina's first encounter with a pizza left her in tears, wondering how she should eat it. With a knife? A fork? Her fingers? Moshe came home crying from P.S. 105 nearly every day because kids would taunt him about his accent, his hand-me-down shirts, his wide, pale, and distinctively Russian face. Alina was terrified by her first welfare assignment. She was ordered to apply at one hundred places for a job in one month's time, long before she had a remote grasp of English. She soon learned that she could just collect business cards from employers without really speaking to them.

Her family navigated their new city in a quivering huddle. The Zukinas clung to each other on their first trip to the Key Food supermarket, afraid that all the store signs looked so much alike, so uniformly red, white, and blue, that they would never find the one they wanted. Before the family's first subway ride, Alina mapped out the route, committing it to memory, as if she were headed on a mission through enemy lines. The family was afraid to take a seat for fear of missing their stop. "My mom finally told me to sit next to a black woman," Alina said. "There was an empty seat. She said, 'If you don't sit there, they'll think you don't like black people.' This is how much we knew."

The family had arrived in the Bronx just after Rudolph Giuliani beat David Dinkins in the race for mayor. Nothing in Giuliani's campaign had forecast his soon-to-be-rapt attention to the welfare situation. More than one million New Yorkers were collecting welfare benefits in 1994, a record number that the mayor set out to slice by half in a few years' time. His first efforts resulted in removing able-bodied single adults from the rolls. But the new mayor expressed a special sympathy for the plight of immigrants, harking back to his own grandfather, Rodolfo, who had come over from Italy in the 1880s looking for work as a tailor. Still, refugees and legal immigrants were subjected to the same strict work rules that Giuliani soon laid out for native-born Americans on public assistance. Two years later, when Congress passed the Personal Responsibility Act, legal immigrants would lose the right to food stamps and cash benefits until they became citizens. They could apply for aid only after they had lived in the United States for five years. President Clinton signed the bill with some reservations, saying he would begin work immediately to restring parts of the safety net for immigrants that the bill destroyed. States like New York, which depended on immigrants to work and replenish its population, were eventually given the option of restoring food stamps and eliminating the five-year waiting period for cash assistance. Giuliani became a strong voice for the immigrants' need for a temporary helping hand.

Such damage control arrived too late for the Zukinas. Alina's first trip to the welfare center at 147th Street in the heart of the South Bronx made the trip to Key Food seem like a pleasant jaunt to the

Odessa Sea. Her family's four months of refugee assistance had run out before any of them could learn English or find a decent job. Like three-quarters of all Russian immigrants nationwide, they then turned to welfare to keep afloat. Outside the Willis Avenue Income Support Center, the streets were filled with an odd mix of busy shoppers and stray dogs, bustling commuters and lonely parking attendants in lean-to shacks. This "Hub" of the South Bronx was once the opulent shopping district for the wealthiest of the newest Jewish immigrants. Nearly a century later the delta of the Grand Concourse was still struggling to repair itself from decades of neglect.

Inside the grim building, Alina and her father lined up with the rest of the applicants. There were no signs to guide them, no workers to ask. Everyone else seemed to know what they were doing—forming two lines to reach one phone drilled into the far wall. Once they reached the phone, the welfare seekers called their assigned caseworker to let them know they were in the building. This arcane practice seemed to serve little purpose other than to add hours to the day, and fray to the nerves.

The first thing that struck the diminutive Russian was that everyone around her was either black or brown. Her only exposure to black people before arriving in New York was from a distance in Kishinev. A handful of African students came every year to study at the university, generally living together in one dorm. Alina and her friends would observe them from afar, giggling at their strangeness. Now she knew the anguish she must have caused them. The only white face in a sea of black and brown welfare applicants, Alina felt the eyes of suspicion on her. She understood English enough to overhear someone wonder out loud whether Russians were taking food out of their mouths— citizens of the United States.

She wondered whether she would ever get used to the racial tensions and her own gnawing prejudice. A few months earlier, in English language class, Alina, the only white student, was having trouble with an essay on Jewish stereotypes. She came up with "all Jews are rich" and then ran out of ideas. Her classmates chimed in with "all Jews are racist" and "all Jews think they are better than everyone else." Alina felt defensive somehow but couldn't summon a coherent argument. "I feel strange there is prejudice among my people against blacks, because we all have the same experiences. I understand it's wrong, but sometimes I feel it too."

Alina left that day with her own case, since she was over eighteen. If she had applied two years later, her welfare clock would have timed out in two years—not nearly enough to help her get through college. But in 1994 single welfare recipients were not yet subjected to the deadline. She would have just enough time to get her bachelor's degree. She enrolled first in Bronx Community College and later transferred to the city's Hunter College for the four-year program. Her father, Boris, left with another case, for the rest of the family. Together their benefits totaled $680 every month plus health benefits and, for a brief period, food stamps. After rent, that left them with five dollars. Without some help from their relatives, the Zukinas could not afford groceries or clothes. Soon after the federal welfare bill passed, they lost their food stamps, never to get them back.

The two Zukinas—father and daughter—were swept up in the nation's most aggressive campaign to move welfare recipients into workfare. Recipients were required to work an assigned city job as soon as their case was opened. By the time Alina and her father got their orders, more than 350,000 people had already filtered through the system. Mayor Giuliani called his workfare system the Work Experience Program, or WEP. He was fond of saying that he was turning "the welfare capital of the nation into the workfare capital of the nation." No other major city came close to having a public-service workforce of this size and scope. It represented the backbone of the mayor's welfare philosophy: give back to get. The social contract. Special language developed around the program. People who worked

WEP jobs were called "Weps." Caseworkers would routinely tell clients that their "free ride was over."

Within a few short months more than thirty thousand Weps were working mostly low-level city jobs in the city's sanitation department, not for wages but in exchange for their benefit check. A work program had existed in the city for decades, but under the new mayor it was enforced and expanded in record numbers. Few were exempt. Orange-vested armies of welfare recipients appeared overnight spearing garbage in the park, mopping courtroom hallways, cleaning municipal building toilets for anywhere from twenty to thirty-five hours a week. Crews of Russian men tromped down each morning to the Staten Island Ferry to mop the decks and clean the facilities.

It was relatively easy to fall off the rolls for noncompliance. Failure to follow the program's myriad rules could mean loss of benefits. Just one workday missed was enough to land a Wep in trouble. In 1995 more than seventy thousand recipients who had been punished took their cases to a state fair hearing. Eighty-five percent of them won back their benefits, an indication that the Human Resources Administration was knocking people off unfairly at a high rate.

WEP jobs offered no sick leave, no wages, no job protection, because they weren't really jobs. When critics complained that WEP jobs rarely, if ever, led to permanent employment, the mayor didn't argue. Instead, he responded that the point was not to train people for a real job but to instill a work ethic for all those on the dole, who had long been stuck in a "cycle of dependency." WEP was not meant as punishment but as inspiration. Workers, the mayor argued, learned important work-world skills: how to get up in the morning, how to be punctual, how to follow directions. It didn't hurt the mayor's case that the city's parks were cleaner than ever, even after twenty thousand municipal sanitation employees were downsized off the payroll.

WEP was a one-size-fits-all kind of policy. It had to be in order to accommodate such huge numbers. Medical assistants with years of experience were sent out on sanitation crews. Mothers with children as young as three years old, then one year old, and later three months old, were sent to scrub municipal offices and bathrooms. Few were exempt from service. The rules lumped together those with a proven track record of work, refugees, asymptomatic diabetics, HIV-infected patients, chronic slackers, and lifelong accountants.

Students like Alina, who couldn't support themselves and earn a college degree at the same time, believed a degree would eventually propel them out of poverty faster than WEP, or anything else. But in order to receive aid, she and others like her would have to busy themselves according to law. Close to twenty thousand students on welfare dropped out of the City University system when the strict rules went into effect. Many said they couldn't fit WEP and parenting and studying into their schedules. Aron Akilov, a teenage immigrant from Uzbekistan, hung himself on a door hook two days after he received a call to WEP work. Workfare duty had meant that the nineteen-year-old man would have to quit his English language classes. Relatives of Aron Akilov said he was depressed about his chances of ever getting ahead if he couldn't master the language first. In the wake of this tragedy, Giuliani defended his policies by urging "people all over the world" to embrace "the American tradition that hard work is at the core of success in America." He added that the "best way out of dependency was not education, but work." The city would not be doing students any favors by underwriting their college education, no matter how poor they were.

This was not an easy concept for Alina to comprehend. In her Soviet world, college students were not allowed to work. Studying was considered close to a sacred activity that would enhance society at large. College grads were guaranteed up to three years of work in their field, albeit for meager recompense, complete with lodging. For both Alina and her father, WEP represented a major hurdle on the path to self-improvement. Alina put in her time at a local social service office that handled Jewish immigrants and other Bronx residents. Boris needed to learn English in order to get a decent job, but there was no time. He clocked in fifty-six hours every two weeks at his WEP job, performing janitorial work for a senior citizens center, in exchange for a welfare check that kept the family well below the poverty line. His English was still worse than his Yiddish, which wasn't good either.

Alina spent nearly as much time in her WEP job, and at welfare offices, as she did in Hunter College classrooms, hoping to earn a degree that would guarantee her family a decent salary for life.

Still, this thin reed of determination juggled her duties with her priorities with a benign sense of "Why not?" Then, after four years of working a steady WEP job for the Bronx Jewish Community Council, the city abruptly cut off her benefits. One of Alina's time cards had apparently been lost between her job and the WEP office. Without questioning her, HRA assumed she hadn't worked and closed her case. When Alina appealed, the welfare agency said she could reopen her welfare case if she agreed to work thirty-five hours a week in a downtown office ninety minutes by subway from her home. The new assignment would not allow her to attend her last classes. She would not have enough time to study.

With graduation only a few credits away, Alina faced the ultimate Hobson's choice. Should she drop her studies, and her dream of becoming a doctor, in order to afford to live?

4

Christine

Patterns

Kristopher Luis Rivera was very nearly born in the city's welfare fraud bureau on November 5, 1997. A biting wind snapped through the North Bronx the day his mother set out to make this appointment in Brooklyn. Christine Marie felt the baby's heavy impatience pressing hard. This child was the reason she had to quit her job as a medical assistant. She knew it was risky to travel on this day. But the irascible thirty-three-year-old believed if she canceled this date to plead her case with the Eligibility Verification Review office, she wouldn't get another.

Christine had had no money of her own for months. She was desperate. She got her back up when she was desperate.

For seven months in a row she had been turned away from her Bronx welfare office. The city believed she had her grandmother's inheritance money in the bank, worth at least fifty thousand dollars. Christine kept insisting that the money had long ago been stolen, "smoked up," and otherwise squandered by her ex, herself, and others. It was a tangled story, scarred by misfortune and naive generosity. She could have been living off the interest. But for better or mostly worse, the money was gone.

The seventh time Christine heard the words "benefits denied," welfare workers had to call the police to drag her out of the Crotona Tremont Center. Christine said she "bugged out," hollering tearful threats. "They kept telling me to shut up and sit down," she said. "I knew I'd be waiting there forever if I did. So I kept yelling, 'There's no money! There's no money!' "

By November, Christine was completely broke and feeding her three kids with occasional cash from the baby's father. She was crazy in love with Luis Lopez, a burly stickball champion with an elegant way about him. But he was feeling squeezed too. He had three kids of his own to care for. She was walking on eggshells around Luis. "I always figured I was one fight away from being in the streets with my kids," she said.

Now, finally, Christine hoped the curse was lifting. The Eligibility Verification Review office was supposed to be the last step before approval for benefits. The mayor had set up this special bureau four years earlier to order up fingerprints of applicants and to run duplicate checks of their personal records. Advocates complained that the office was just an expensive nuisance designed to find more ways to trap people and bounce them from the rolls. The office went over the same documents already reviewed by the welfare office. The city claimed it was necessary to double-check in order to ensure that only the neediest applied.

Amid the vast constellation of dingy city welfare offices, the EVR was one of the least appealing. It was an unpleasant place for paper pushing, let alone birthing. On any given day garbage bags were piled high outside its unmarked doors on Jay Street in downtown Brooklyn.

Steel window bars separated applicants from the city's fraud workers, leaving most clients with the impression that they had just narrowly escaped a jail sentence.

Christine wasn't thinking about any of that when she decided to take the long train ride from the northeast Bronx to downtown Brooklyn. She toted a bag of documents, waiting for her name to be called. Waves of labor pains sucked at her breath. Christine felt a warm trickle, and then a river of water rushing between her legs onto the plastic chair, onto the cracked linoleum floor. The woman next to her began to yelp. Christine panicked. Christine grabbed her bag and tried to flee. All she could think of was to get herself to Albert Einstein Hospital, at least two hours north in the Bronx. She needed her doctor. Only her doctor knew what to do. Christine's was a high-risk pregnancy. The fetus was surrounded by potentially lethal strands that used to be its amniotic sac protecting it from the placenta. The sac had ruptured weeks before, forming bands that could cut the baby, maybe even sever a limb or two.

Police tried to stop her from leaving the welfare office.

"You're not going to make it. We'll call an ambulance," the officer said.

"I can make it. I have to make it," Christine replied. She hobbled off to the F train at Borough Hall, officers at her side, trying to talk her out of this foolish venture.

"Don't do this. The pains are coming once a minute," the police officer pleaded.

"I need to get to the Bronx," Christine repeated, between contractions.

Finally, on the sidewalk at the top of the subway stairs, the Puerto Rican mother gave in to the pain. She heaved herself into the patrol car. It sped off to the nearest hospital. Kristopher was born a half-hour later at Long Island College Hospital, scarred not by the amniotic bands but by his mother's heroin habit.

• • •

When Kristopher was an eighteen-month-old toddler, Christine bundled into the consulting room at the Jackson Avenue Family Shelter one May afternoon, limping from a lunge down the hall to grab the baby. The stroller was draped with plastic bags, diaper bags, shoulder bags. The mother nudged its wheels—once, hooked, twice, hooked, three times, free—through the door.

She navigated a table leg as if in a hypnotic haze, her curly hair stuck with sweat to her back. A heart-shaped locket flipped open, the photos it once held long gone. She glanced up with a bright smile. The charismatic woman behind a life of distractions emerged briefly to greet a newcomer. Then, back to the business of chaos. Kristopher whipped out of the stroller. The toddler promptly climbed on top of the table in his miniature work boots and bounded from one end to the other, until he collapsed in her arms for a voluntary time-out. Seven-year-old Dyanna, another of her four children, skipped in to ask whether she could play with a friend, her braids sticking out playfully in three directions. She threw her lithe arms around her mom's neck for an impromptu hug, then danced out the door.

Christine's defiance was spirited, jarring, in a pleasant way. She was rough-hewn, but not angry. It was the same stubborn trait that had caused her to leave the house that day in labor. In spite of a string of questionable decisions along the way, she had charged pell-mell through life, getting along by her wits and her warmth. She was a bright chatterbox, innately trusting. She didn't fight her circumstances so much as try to climb out of them, as if all she needed was the right ladder. "I'm going to get out of this," Christine said, with conviction, while Kristopher used her lap as a jungle gym, his little fingers yanking on her nose. "I'm going to get an apartment. I'm going to go to school. I'm going to get off welfare. I'm going to get my job back." Step by step. "I'm clean almost every day now." It was her survival mantra. She would repeat it often in the months to come.

If Christine were to succeed in putting all these life puzzle pieces together before her welfare deadline ran out, she would be defying boxes of national statistics. Mothers in Christine's situation posed the most exasperating challenge for those who promised to end welfare by the millennium. Hank Orenstein saw hundreds of women walk in and out of his shelter who faced a numbing array of barriers to work. They were single mothers who tended to have minimal education, on-and-off work experience, and more than two children. For reasons experts were still debating, a disproportionate number were Hispanic—an ethnic group left behind in the great welfare exodus. Social scientists had identified specific risk factors for women, conditions that would sabotage any economic progress: depression, drugs, violence at home. Most of those left on the rolls were damaged in soul and body, people for whom public sympathy grew dimmer by the year. Christine could check off all three risk factors. She had grown up in a South Bronx fractured by seventies' drugs, eighties' violence, and nineties' poverty. Her alcoholic mother died violently when Christine was a teenager. From the age of three until she was thirteen, Christine said she was sexually molested by the man she then believed was a close relative. The odds of success, as academics calculated them, were stacked against her.

Caseworkers had long recognized that a large percentage of the homeless and welfare population were abused as children, sexually or physically. About one-quarter were traditionally believed to be vulnerable to some form of mental illness or substance abuse, according to the federal Department of Health. Now the percentage was much higher as the hard-core and hard to place were left on the rolls and the more easily employable left welfare for work.

Christine hoped she would be the one to buck the odds. She'd always managed to keep her own family together in her own way.

In recent years it had looked as if her life was veering safely away from the endless ordeals toward hints of serenity. But then two inevitabilities collided: her drug addiction and welfare reform. It was impossible to sort out which caused what. Did the public policy trigger her weakness for drugs? Did the addiction overwhelm everything in her life? The only clear thing was that Christine's drug habits spiraled out of control and her family was torn apart shortly after she entered the system of strict welfare regulations and shelter rules. Deadlines, work rules, curfews, forms, orders . . . Christine didn't do well with any of these. She didn't think in those terms. She balked. She scheduled conflicting appointments. She lost questionnaires. The new welfare rules no longer forgave her these trespasses.

"I have everything to live for," she said at one point, in a heartfelt narrative of her predicament, "and not much to live it with. I was gonna say I don't know how it happened like this, but I guess I do. Some of it is my fault. But not all of it."

Christine came from a long line of proud Puerto Rican women whose best intentions were often eclipsed by events more powerful than their ability to fight them. For years after World War II, Christine's great-grandmother plotted her escape from Puerto Rico. Mercedes Guzmán cobbled together her savings, hiding them from her husband, as she made plans to find work and set up a new life in New York City. Her husband would have killed her if he had found out what she was doing. He was a jealous man who wouldn't let Mercedes have her own friends or wear makeup or nice dresses. When she'd finally saved enough for passage, she packed up her baby, Marie Carbonell, and went to mainland America to forge her own way.

The lore in Christine's family held that Great-Grandma Guzmán paid two dollars a week in rent from her seven-dollar-a-week welfare check until she finally got a job. She came with a great wave of Puerto

Rican migrants fleeing the overcrowded island, tired of the sugar- and molasses-based economy. There was a boom in postwar industrial work around New York and no shortage of housing. Many immigrants edged away from the barrios in Brooklyn and East Harlem, popular before the war, and settled instead in the South Bronx.

That's where baby Marie grew up with a loving mom, and to a life of labor. Her first job was folding clothes in a laundry. Then she took a factory job, a dry cleaners' job, and finally a civil servant's position with the post office. Marie held on to the post office job until she retired, collecting enough bonds and pension money to live comfortably. When Marie died, she left a sizeable inheritance to her eldest grandchild, Christine. It was money that should have saved the bright girl from a life of poverty and poor choices in men.

Christine always carried a crinkled photo of her mother in her wallet. In the one-inch-square snapshot, Diana Vidal is a handsome woman, on the thin side, wearing a plaid summer dress. She is sitting on her couch, leaning on a man—the love of her life. This was the man Christine said routinely molested her from the age of three until she turned thirteen, the age when the little girl finally realized she could speak up and defend herself. "He used to call me beautiful, his little princess. He told me not to tell," Christine recalled, her bright eyes flashing anger. "Nobody talked to me about sexual abuse. I didn't know what it was. No one told me it was wrong." When she finally threatened him with arrest, he stopped. The young girl never did press charges or seek medical or psychiatric help. She couldn't even tell her mother. "My mom really loved him," Christine said. "She couldn't have handled it."

Diana went on and off welfare as she raised Christine and her younger brother. In between, she worked as a medical assistant, a teacher, and later a clerk in a children's clothing store. When the couple broke up after a decade, Diana started drinking heavily. "She nearly lost her mind," Christine remembered. "Once, she pushed furniture against the door, barricading herself in the apartment, screaming, crying, lashing out with knives at people."

It was about this time, when Christine turned fifteen, that a relative introduced her to the sedative and seductive powers of marijuana, and later the deadly euphoria of crack. A cornucopia of drugs followed, from angel dust to her ultimate nemesis, heroin. She became a sometime student in Adlai Stevenson High School in the Bronx, bored with everything except woodworking, and barely above water with her grades. The same year she got pregnant for the first time and had an abortion. Her boyfriend made her feel so guilty about it that when she got pregnant again the next year, she kept the baby. Mark Anthony Cruz was born in 1981. Christine was sixteen. Somehow she managed to graduate from high school and later went on to get a data processing certificate, one of her precious achievements. Mark's father died of leukemia two years later.

That same year, 1983, marked another tragedy, the worst catastrophe yet. It was three days before Christmas. Diana Vidal was watching her grandson, Mark, a two-year-old live wire by then. She had been drinking, which was nothing new. A loud argument erupted between Diana and the man who'd been living with her for a while in their apartment on Morrison Avenue near the Grand Concourse. Christine never liked that man. He was a bellicose drug addict, always picking fights, popping heroin, stealing her mother's rent money. She remembered seeing him once walking around the house with a needle hanging out of his arm.

Neighbors heard their two voices rising in pitch. Then, in a sudden gasp, Diana fell out of the fifth-floor window to her death.

Christine's friends believed it was suicide. The police dismissed the fall as a drunken accident. Christine always thought her mother was murdered. The day after it happened, baby Mark kept saying,

"*Buelo* push, *Buelo* push." "Old man push, old man push," she remembered. Christine was too young and too numb to pursue prosecution. But she never truly recovered.

"I've been on my own since my mother died," Christine said. "I don't want people ordering me around. I can't take it when they order me around. I will do what I need to do. I do for myself."

The trauma of losing her mother this way—suddenly, violently—left Christine almost stuck in time. At thirty-three, she was still very much an ingenuous teenager, hankering for a good time, searching for ways, risky and riskier, to deaden her pain. "She was a good mother," Christine said, "when she wasn't drinking."

Since she was a minor, Christine inherited her mother's welfare case at age seventeen. She also inherited many of her patterns—losing love, work, and children to addiction and violent men. Welfare was always a necessary annoyance in both their lives, resorted to when pregnancies and addiction eclipsed their ability to cope.

When she was an older teen, Christine met Francis Geovaine Munnings, a rough-edged high school dropout from the neighborhood, one year her senior. He worked on and off as a porter at the Hunts Point Terminal Market. It wasn't long before his penchant for coke, crack, heroin, alcohol, and violence took over Christine's young life. Munnings beat Christine while she was pregnant with their first child, Monica, and later with Dyanna—his two daughters. She quit her clerical job and went on welfare to take care of the babies. He squandered her grandmother's inheritance, buying drugs and cars and more drugs. Christine did her share of squandering too.

The worst came when Christine discovered that Francis had molested a five-year-old child. She knew he had his troubles. He'd already spent several years in jail for selling drugs. He was on probation for stealing a Chevy van. But Christine couldn't believe Francis actually would assault a little girl. The doctors' testimony convinced her it was true. Christine helped turn Francis in to the police. He was sent to jail in 1992 for committing sexual abuse and endangering the welfare of a child. "I swore I wouldn't put up with the same stuff my mother put up with," Christine said. "But I did, for ten years. Don't ask me why."

In many ways it was a minor miracle that Christine and her children emerged from this spiraling trauma into a functional existence. Psychic damage had taken its toll, untreated, unrecognized. But even so, Christine was able to overcome it on her own, for a while at least. In fact, Christine was never happier than she was in the years leading up to Kristopher's birth. She had a job she liked and an apartment in a quiet residential section of the northeast Bronx, away from that borough's hardscrabble drug streets. She was living with a man she loved deeply, Kristopher's father, Luis Lopez, a dignified bear of a man with a tasteful earring and melodious voice. Christine always called him her "husband," though they never married. For four consecutive years they lived in a drug-free home with her children. She was not on welfare. This was a real life, Christine thought. She had made it. She had survived the worst beating this world was going to give her. Her ex was locked up. Her relative was no longer around blowing crack in her face when she was trying to clean up. She missed her mother, terribly at times. But still, this was as free as she'd ever been.

Luis set up his own business in his Cruger Avenue apartment as a freelance Spanish-English interpreter for clients in the courts, at fair hearings, and involved with Social Security. When business was slow, the forty-five-year-old fixed cars or sold T-shirts and tapes. When the U.S. census needed surveyors, he signed up for the job. He had three other kids from another marriage to support. Christine credited him with helping her learn how to get along without fighting and with keeping her bloodstream clear of drugs.

Christine was crazy for Luis. "I get all tingly whenever I see him," she said. He took Christine to Puerto Rico for the national stickball championship tournament. It was the first time she'd ever traveled outside of the city, even outside of the Bronx. Memories of the island's white sand and azure ocean kept her imagination active years later. He promised her another trip, to Seattle, to visit his family. She looked forward to it like an adventure-starved teenager.

"He's a good man. Very generous."

Christine worked as a medical assistant at the Hunts Point multi-service health clinic, located on one of the nation's most poverty-stricken peninsulas. For six dollars an hour, she dealt with AIDS and mental patients, doing paperwork, taking blood. There were always openings for that kind of work, Christine said. "Most people are scared of getting AIDS and hepatitis from these patients," she said. "They leave it to people like me. I don't mind. I'm good with patients." She couldn't earn nearly enough to sustain a family of five in New York City. But with Luis helping out, she could make it. And she was hoping to attend classes to get more training in phlebotomy, the science of taking blood.

Then Christine got pregnant with her fourth child, Kristopher. She didn't think she could conceive. She'd had two miscarriages in recent months and one abortion. The pregnancy was tough. She had to quit work and stay home. A familiar depression began gnawing at her. Her past unresolved and untreated anguish welled up like a nightmare. She was alone. She hated being alone. Alone, she felt helpless against the current of despair. She started using heroin again, to scrape the edge off the pain. She hated herself for it.

"I wanted to quit, but I couldn't. I couldn't go cold turkey, because the baby could die," she said. The term "cold turkey" refers to the rash of goose pimples that spreads across the skin of the addict long over-due for a fix. "Every time I needed a fix, I could feel the baby knot up in my stomach, withdrawing. I couldn't believe I was doing this. It was horrible. Horrible. I hated it when people did that to their babies. And now here I was. It was the worst thing I ever did."

Christine took to buying synthetic methadone on the streets, as a slightly less toxic brew. "I cried every day because I hated meth, but it wasn't as bad as heroin. I got dependent on it too. I had real bad guilt. I was so scared. What am I doing to my baby?"

Day after day she fretted at home, trying to figure her way out of this mess. She wanted to ask Luis for help, but he'd always told her, "Do drugs and you're out of here." She knew Luis would be disgusted with her once he found out. Finally, at seven months pregnant, with no options left, she confessed. Luis blew up but didn't throw her out. He couldn't toss a woman pregnant with his child into the streets. Instead, the two fixed up a nursery with a crib to make a good showing for child welfare workers after the baby was born. Since Kristopher would inevitably be born with dope in his system, Christine would have a child welfare case with the city as soon as he arrived. That meant regular visits from the social workers. The burden of proof would be on Christine to prove herself fit to be his mother.

Christine couldn't stop. The fetus was addicted by now. Withdrawal could kill it. She couldn't stop stealing from Luis to pay for her habit. She knew she could lose Kristopher and Luis both. But addiction was not rational. It was a progressive disease. She convinced herself that she was in control. She could be mainlining, she told herself, but she adamantly avoided needles. That was a small source of pride. She was sniffing heroin so regularly that she needed the drug to feel normal—not high, but simply pain-free. Without it, she was achy, sweaty. She still had no welfare case, no money of her own. The welfare office had rejected her four times in four months. Then, one month before Kristopher's first birthday, Christine stole one of Luis's big paychecks. "She was just too repetitious. I flipped out," Luis said later, as we waited in family court for one of several hearings.

"None of my children were ever born with drugs. He tested positive right away," Luis remembered, blanching at the two-year-old memory. Kristopher went through withdrawal and had to be pumped with phenobarbital to prevent seizures. "It's inexcusable to do that to a child," he said.

Luis told Christine she had to leave.

Christine took her four children to the Emergency Assistance Unit on East 151st Street, the same portal to welfare that Brenda Fields had entered two years earlier with her children. She noticed the Bronx Supreme Courthouse sitting white and majestic up the hill to the north, above Yankee Stadium. Looming even closer down the bluff were the vertical correctional faciliies built by welfare workers during the Depression. It was where prisoners were held before arraignment, sometimes fifty to a cell. The Bronx office of the Administration for Children's Services was right around the corner. There city social workers arranged foster care for children removed from their homes. Her world was closing in on her. She found some bare spots on the floor for her family to try to sleep for the night.

Part II

Work

If any would not work, neither should he eat.

—2 Thessalonians 3:10

Work, work, work, is the main thing.

—Abraham Lincoln, 1860

5

La Guardia's Legacy

Rudolph William Louis Giuliani was born to first-generation Italian immigrants in East Flatbush, Brooklyn, on May 28, 1944. Nine days later, U.S. troops stormed the beaches of Normandy on D-Day. Europe would soon begin its painful recovery from years of Nazi atrocities, and New York City would finally vault out of its own financial depression. But no global or local event could overshadow the celebrations of the extended Giuliani family over the arrival of Harold and Helen Giuliani's only child. The newborn son was considered something of a miracle to his parents. Already in their mid-thirties, Harold and Helen had reluctantly adjusted to the prospect of a future without children. From birth, the child was protected by his admiring family, educated in some of the city's finest Catholic schools, and groomed for hard work and tough competition—the Old World way.

Some fifty years earlier, Rudy's namesake and fraternal grandfather, Rodolfo Giuliani, had set his own courageous example. The seventeen-year-old tailor made the journey from Tuscany to Ellis Island, joining the waves of Europe's hopeful emigrants at the turn of the century. Once in New York, Rodolfo met and married another Italian immigrant, Evangelina, who spent most of her life sewing in the Garment District sweatshops, raising their five children when she could. Theirs was a storied life of toil and sacrifice, buoyed by the American promise of opportunity and Rodolfo's particular affinity for Italian opera. Secular devotion to the work ethic would eventually pay dividends—both spiritual and material—to their children, and to their children's children. Parental sacrifice equaled higher education and a prosperous life for the offspring. And as for Rodolfo's passion for opera, it provided a simple, conceptual retreat into the soulful hills of his heritage, one that his grandson would later share.

The seamless path from work to prosperity took a jagged turn when Rodolfo and Evangelina's firstborn son came of age. Parents and child were products of different countries, different cultures. Harold Giuliani dropped out before completing his high school degree. He took odd jobs in odd intervals as an off-the-books janitor, a part-time plumber's helper, a sometime bartender. *Village Voice* reporter Wayne Barrett revealed in his book *Rudy: An Investigative Biography* that Harold drifted into Brooklyn's seamy underground, working a stint as a collection agent for his uncle's mobbed-up loan shark operation. Once, when he was twenty-six and jobless, a desperate Harold robbed a milkman at gunpoint, a crime that sent him away to Sing Sing for more than a year. Throughout his checkered life, the tailor's eldest child was never abandoned by his extended family. Welfare was not considered. Generous relatives were always there to support Harold's family financially during his bouts of unemployment. And his only son, Rudolph, never strayed from publicly admiring his father's spunky aggression. Whenever Mayor Giuliani referred to his father on the campaign trail or in a welfare-to-work speech, it was to celebrate Harold's values, his lessons of hard work that were bestowed upon his son.

If Rudy recognized the hazy outlines of his father's morally ambiguous life, he did not feel compelled to reconcile them with his own stark views on human behavior.

Later in his political career, Rudy Giuliani frequently summoned the brightest parts of his immigrant family's story to make the larger political point. The greatness of New York was built on the sweat and integrity of people like Rodolfo the tailor and Evangelina the seamstress, the mayor would say. These were citizens who worked for every nickel in their pockets. They lived out the end of their lives in a city rescued by federal welfare programs, yet they did not tether themselves to the degradation of public relief. As mayor, Rodolfo's grandson hoped to return New York to the mythic era when hard work was a prerequisite for pride and strength of character. Nurturing these roots on a personal level, Rudolph developed a love for the grand arias of *Figaro* and *La Traviata* with a fervor equal to his grandfather's.

Fiorello La Guardia was in his third term as the city's most beloved mayor when Rudy was learning how to button his toddler-sized Yankee uniform. The "Little Flower," who read the Sunday comics over the radio during a newspaper strike, had by then accomplished some monumental feats in New York politics. He had wrenched political control of the city's voting machines from the entrenched corruption of Tammany Hall. He had rescued the city's budget from J. P. Morgan and near-certain bankruptcy. The pint-sized Republican with big government ideas snagged 20 percent of the New Deal contracts granted under President Franklin Delano Roosevelt's programs. With this federal bounty, La Guardia put thousands of unemployed New

York City men to work building roads, schools, bridges, subways, parks, and hospitals.

The city was climbing out from under the suffocating gloom of the Depression, and its citizens were grateful to the compassionate mayor who lit the way. He understood the plight of the little guy. He knew New York's weakest links could become its greatest assets, somewhere down the pot-holed road. The Columbia University historian Thomas Kessner wrote that La Guardia "fought a lonely fight trying to persuade a prosperity-drunk America that the poor deserved its attention, that a nation could not substitute a cash register for a heart and keep its moral balance."

Rudy Giuliani grew up admiring the rumpled visionary. La Guardia's popularity was a source of pride in the Giuliani family. This half-Italian, half-Jew with the mellifluous Old World name had no connections to old American money or older Protestant privilege. Yet he had managed to unite a crumbling city in order to save it. And he did so with a zeal for sorting out right from wrong and a sense of human compassion. "What is needed, is government with a heart," La Guardia said. "I will choose to help the suffering."

In 1989, when Giuliani first announced his intention to run for mayor, the high-profile U.S. attorney chose the same clubhouse used by his idol fifty-six years earlier. Giuliani would happily wrap his campaign in the modest trappings of his predecessor, the bantam mayor. Taking the podium at Manhattan's Metropolitan Republican Club on the Upper East Side, Giuliani spelled out his La Guardia-esque plan to ease the city's suffering, to kick out the "old, tired, corrupt administration." Giuliani chastised his opponents' tendency to belittle the poor, saying, "Each time the administration attacks those less fortunate by exaggeration, and cruel characterizations, New York loses a bit of its soul." A few weeks later the candidate spelled out his desire to ease crushing poverty and to provide treatment for drug addicts. Government's most noble purpose, he said, was to help people conquer the difficulties and pains of life. "All of my past draws me to take on this challenge," said the candidate, "to restore the city of my grandparents and parents, of my relatives and my friends, and to offer New Yorkers hope for the future once again."

Giuliani lost his mayoral bid that year to Harlem Democrat David Dinkins. But four years later, in 1993, he reversed the defeat, replacing the first black mayor in the history of New York City. In Giuliani's first week in office, the newly elected mayor paid homage to his political muse by attending a performance of the 1959 musical *Fiorello,* signifying the kind of fusion government between liberals and Republicans he hoped to construct.

But the nobility of La Guardia's large social programs had lost much of its political appeal for Giuliani's party long before he took office. Giuliani had come to view the ballooning city welfare rolls as the cause of New Yorkers' poverty rather than as its temporary solution. From this new vantage point, more than one million New Yorkers were trapped in an unworkable benefits system that required too little of them in exchange for a small check. And the numbers were growing. La Guardia's "government with a heart" in the thirties and forties had become so expensive, so unwieldy, so lax, that it required a respirator in the nineties.

Instead of choosing to "ease the crush of poverty" as his first order of government business, Giuliani decided to focus instead on reducing crime by half during his tenure. Law and order emerged as his primary issue. Cutting government spending became his primary message. The new mayor now believed that social order was compromised when the poor were offered benefits with no obligations attached. Balance could be restored if the vast majority of the welfare population was required to work for the city in exchange for rent and food. Strict application requirements and stricter work rules would result, Giuliani believed, in fewer government dependents, fewer crimes, and a larger budget surplus.

Thus, compassion came to mean something different for Giuliani than it did for La Guardia. A truly compassionate government, he argued, would free its citizens from the yoke of public assistance, not foster dependence on its free-spending generosity. Recipients would be encouraged to strive for self-sufficiency. "It takes more compassion to help people help themselves," Giuliani told an audience in North Carolina in 1998, "than it does just to throw money at them." Work in exchange for benefits became a City Hall mantra. "For every right there is an obligation," the mayor would often say. "For every benefit there is a duty."

Such seesaw opinions over the real purpose of public aid were as old as the nation itself. New York City opened one of America's first poorhouses in 1736 in order to isolate the poor while simultaneously offering shelter. Modeled after England's Victorian institutions, poorhouses were designed to enforce behavior codes and to keep residents from applying for relief. Most citizens believed that even these relatively spartan services encouraged sloth and sinfulness. Colonial versions of workfare ranged from picking "oakum" to weaving cloth. Handouts of food or coal (rarely both in the same week) were the early equivalents of food stamps. According to the social historian Michael Katz, critics back then argued that luxuries such as tea and sugar should be excluded from the one-dollar potato and flour rations. Others worried that the poor would sell the coal to buy alcohol or other morally questionable goods. By the end of the nineteenth century, the state of poorhouses had deteriorated into something scandalous. Fear of ending up in such an establishment became the most effective public policy incentive, stoking the public work ethic.

The next wave of charity, in the nineteenth century, focused on the children of the poor. Protestant and Catholic rescue services known as "child savers" gradually came into prominence. Missionaries from all faiths pushed for child labor laws, better health conditions, and compulsory schooling. Many missions set about removing children from the ghastly poorhouses or indentured servitude, sending the youngsters to orphanages or other institutions. Beginning in the 1850s, the Children's Aid Society rescued abandoned street urchins from New York City's grates and gutters and shipped them on "orphan trains" to be adopted by farmers and townspeople in the Midwest and beyond. After decades of forced adoptions and institutional solutions, the poverty debate then returned full circle to the value of preserving families. Mothers were brought back into the policy discussion at the turn of the century. Widows were seen as particularly worthy of government aid. Several states began distributing "mother's pensions," sometimes called "widow's pensions," to help them sustain their families without working outside the home. This was the beginning of the entitlement era in social welfare, a prequel to the New Deal's Aid to Dependent Children.

Until the Great Depression, the federal government had kept itself conspicuously absent from matters of local food and poverty relief. But the mass unemployment and starvation that followed Wall Street's collapse in the 1930s ignited a sense of urgency for national solutions. Conditions were shameful and dangerous. A shantytown nicknamed "Hoover Valley" popped up in the abandoned basin of New York City's Central Park Reservoir. Rent riots erupted in the mostly black neighborhoods of Chicago after scores of men attempted to return an evicted family's furniture from the streets back into an apartment. Waves of unemployed workers stormed the cities' municipal offices and abandoned buildings. These public crises focused the public eye on government's responsibility to its poorest families.

President Franklin Delano Roosevelt responded to the looming social chaos by creating the first federal relief system: Social Security for the elderly and disabled, large-scale work projects for the unemployed, and direct Aid to Dependent Children. These programs would become the cornerstones of America's welfare state, with work as the driving mission of nearly every project. Roosevelt aimed to build programs "that put their faith in the forgotten man at the bottom of the economic pyramid." Still, the far-thinking president never intended for benefits to become a way of life. Anything more than temporary aid would become a "narcotic, a subtle destroyer of the human spirit," the president said. He didn't live long enough to witness the dynamic social upheaval to come.

Not far from the federal offices where Giuliani was launching his career as a U.S. prosecutor, the welfare rights movement was fomenting in the 1960s, rattling the foundations of America's welfare system. What began as a small program on Manhattan's Lower East Side to curb teen violence would grow during the decade into a revolutionary attempt to eliminate poverty nationwide. The idea put forward by two Columbia University social work professors, Richard Cloward and Francis Fox Piven, was to "flood the rolls" and "bankrupt the cities" in an effort to force an even more radical federal solution to poverty. The activist scholars hoped the new army of eligible recipients would cause such a strain on local welfare offices that they would inspire Congress to adopt a guaranteed national income. The movement came breathlessly close to achieving its goal.

When Cloward opened up Mobilization For Youth, the office was immediately flooded, not with teens but with indigent families seeking food, shelter, and clothing for their kids. Most of them were recent black migrants from the South who had fled racial terror and the injustices of sharecropping. Nearly everyone who came to the Lower East Side office was eligible for welfare benefits, but only one-third had known to apply. So the MFY staff began doing what it could, cajoling the system to provide benefits for these second-generation casualties of slavery. The welfare lexicon changed for recipients, from a passive request for assistance to an aggressive demand for entitlements. Cloward and Piven figured these desperate families were not simply eligible for but "entitled" to receive assistance, and everything else the law allowed. By 1966 New York's rolls had doubled to half a million recipients—Depression-era numbers. Other cities in Ohio, Pennsylvania, and California experienced the same precipitous rise. In a decade's time the nation's welfare dockets nearly tripled in size. And the new complexion of welfare was gradually transforming from white war widow to black unmarried female with children.

Thirty years later Giuliani blamed this movement, which later inspired Lyndon Johnson's War on Poverty, for nearly everything that was wrong with welfare in the city he inherited. "From 1960 to 1994, the work ethic was under attack in New York City," Giuliani said in a major welfare address in the late nineties. City leaders had romanticized welfare, he said, creating a "philosophy of dependency" that robbed the poor of their ambition. "It's a terrible thing to do to people."

The first New York City Human Resources Administration was set up in the midst of this storm in 1965. Commissioner Mitchell Ginsberg instituted procedures that would make it easier for people to collect benefits. He simplified the application form. He required minimal proof of identity or need. He replaced social workers who managed cases with eligibility clerks who simply processed applications. He stopped the often humiliating visits to applicants' homes by social workers who would search in closets and under beds for signs of a male wage earner. Some welfare clerks from the mid-sixties remembered rows of school buses parked outside their offices, engines running during the cold winter months. "People would start lining up outside the Hamilton welfare center at five, six o'clock in the morning. They would climb in the buses to keep warm," said Gwen Richardson, hired fresh out of high school as a welfare clerk. The flood of needy people collecting aid was so overwhelming that the *New York Daily News* dubbed the new commissioner "Come-and-Get-It Ginsberg."

The specter of so many unmarried women of color with children lining up to get something for nothing rankled many in Washington. The idea that the recipients should work off their checks began to percolate in Congress in the late sixties. Senator Abraham Ribicoff of Connecticut argued that recipients could easily be handed "a stick, a broomstick with a spike to pick up a piece of paper," to work off their check. It would require little training, he reasoned, and would result in cleaner streets. But New York was not ready for it. Lawmakers believed that any mention of such forced labor for this population smacked of slavery. New York City's Republican Mayor John Lindsay called the idea "coercion that could alienate the poor." New York's Democratic Senator Robert Kennedy said this concept represented a "punitive attitude . . . reminiscent of medieval poor laws."

Despite its liberal critics, a welfare backlash picked up steam in the seventies and eighties. Presidents Richard Nixon and later Ronald Reagan launched effective public relations assaults against "welfare queens," a relatively small number of women who scammed the public assistance system to support their steak and Cadillac habits. States eventually set up welfare investigation offices. Legislators and bureaucrats became focused on fraud and error rates. In the seventies, New York's error rate was about 27 percent, meaning that 27 percent of its welfare recipients were later found to be ineligible. By the mid-eighties, that number was down to 6 percent. New York State wanted it to be even lower. As a sign of the toughening times, benefits were frozen at levels that left families well below the poverty level. Finally, work was required in the early seventies—at first only for single adult recipients seeking assistance. By the late eighties, single mothers with children over six years old were also subject to work requirements. Welfare, always an unpopular program, was suffering a public relations meltdown.

When Giuliani took office in January 1994, more than one in eight New York City residents was receiving public assistance. Still, it would take nearly a year before he turned his attention to them. The new mayor decided first to address what he called the quality of life for the majority of citizens. He set out to clean the streets of peddlers, beer drinkers, and petty criminals. He enhanced the police budget. By his second year Giuliani quietly predicted a small rise in the welfare population from 1.1 to 1.3 million. But then a Westchester County welfare program in upstate New York caught his attention. Its aggressive plan identified fraudulent claims up front and required every recipient to work on some form of a municipal crew. Giuliani came to believe that simple rules changes and enforcement might eliminate thousands from the rolls and revitalize the work ethos with minimal government effort.

Giuliani went after the perceived welfare cheats with his usual dogged persistence. One year before Congress passed its Personal Responsibility and Work Opportunity Reconciliation Act, the former prosecutor set up a new anti-fraud bureau staffed with fifteen hundred clerks and former cops charged with purging the welfare rolls of ineligible claims. He approved a multimillion-dollar electronic fingerprinting program. Workers ran credit checks and searched for unreported bank accounts, stock options, burial plots. They scanned the system to make sure the applicant was not applying in more than one neighborhood or hiding a hefty stash of cash. Investigators wearing FEDS (front-end detection system) badges scheduled home visits to verify addresses. They searched law enforcement records to make sure applicants were not wanted by police. A 1997 state study found that the expensive finger imaging did little to deter fraud, particularly when so many other rigid rules were already pushing people off the rolls. And since far fewer than 1 percent of applicants were ever prosecuted for fraud per year, the $13 million system seemed to work far better at deflecting new recipients than catching cheaters.

Still, the federal government rewarded the measures for their deterrent effect. In May 1999, HRA received Washington's "Hammer" medal for its work in eliminating more than 1,000 fraudulent applicants from its disability rolls, saving a total of $15 million. In November 2000, HRA announced that more than 7,000 welfare cases were closed after fraud workers discovered that the applicants had outstanding felony warrants. Another 1,250 fleeing felons were arrested when they tried to apply for welfare.

These were promising developments for the new administration. Yet Giuliani understood that warding off cheats and felons had its limits as a comprehensive welfare reform tool. Such vigilance helped to police the raw edges of the welfare world, but it did little to move bigger numbers off the dole and into the work world. To accomplish this goal, Giuliani believed that the poor required more structure and order in their chaotic lives. The Work Experience Program was born. Armies of welfare recipients were dispatched to sanitation and parks department crews. WEP workers were required to follow work rules or risk losing their benefits. The city would be cleaner. The poor would not necessarily be learning marketable work skills. But they would be learning essential work ethic values. And the rigid rules regulating the workfare crews would result in reduced caseloads, thus saving the city millions.

Adults with no children were the first group to be called in for WEP assignments. Single mothers with children as young as three months old soon followed. Those attending the City University of New York were not exempted from work in order to study. Close to twenty thousand dropped out in a few years' time. About thirty-five thousand WEP workers at one time would put in twenty to thirty-five hours a week cleaning municipal buildings, scrubbing subways, erasing graffiti, and picking up trash from the city's parks and streets with spiked sticks, just as Senator Ribicoff had once proposed. No other city in America had launched such an industrial-sized workfare program so quickly.

The mayor's give-to-get policy began to work quickly, by way of attrition. From 1995 to 1996, the welfare rolls dropped 150,000 recipients. By the next year, 350,000 had fled the rolls. The rapid decline surprised nearly everyone, most of all the mayor. No one could precisely measure how much the booming national economy played a part. But the rolls continued their staggering decline, with no immediate, dire results. The crime rate continued to fall. Families did not appear to be seeking shelter on sidewalk grates or park benches. The homeless had not yet taken over City Hall's steps. "If a quarter-million people removed from the rolls had no other place to go, you'd see homelessness increase," said Richard Schwartz, the city's workfare architect, in a 1997 interview. "You'd see the crime rate go up. And that hasn't happened." Schwartz left the Giuliani administration that same year to start Opportunity America, a for-profit welfare-to-work business.

The city could not produce data to show how many of those who left the rolls had found self-sustaining jobs. Nor did it track how many WEP workers graduated into private-sector jobs after their workfare requirement was completed. Some were clearly getting private-sector jobs; others were transferred to the federal SSI rolls. Still others disappeared into the untrackable underground economy in New York City—sewing sweatshops, street vending, exotic dancing, grocery delivery, baby-sitting, housekeeping. Giuliani embraced the rapid decline as a sign that "work and the values of work" had made a revolutionary comeback in New York City.

Advocates for welfare recipients criticized WEP for offering make-work jobs for people who needed education and training instead. They accused it of being a forced-labor system designed to humiliate and hassle minorities rather than to help them gain marketable skills. A local food advocacy group found that only one-tenth of 1 percent of WEP workers left the program for a private-sector job in its first year. The *New York Times* found that 69 percent left the welfare rolls in the early years not because they had seized an outside job opportunity but because they violated one of the work rules associated with WEP.

Stocked with anecdotal material, critics argued that WEP was also dangerous. A forty-year-old man collapsed on the street from heat exposure one summer day after working his WEP shift picking up garbage around Gracie Mansion, the mayor's residence. A fifty-year-old woman who had lost her previous job because of a heart condition was forced by the city to work a clerical job in exchange for her welfare check. During a lunch break from her WEP job, she collapsed and died of a heart attack on the Coney Island boardwalk. Another young mother died of a brain tumor because she was afraid of missing WEP hours to wait for a doctor's appointment, thereby jeopardizing her welfare benefits.

Giuliani argued right back at them. He said these incidents were used unfairly by a liberal establishment intent on propping up the old dependency culture. He had his own anecdotes, like Mercedes Ovalles, who was hired full-time by the city's Division of AIDS Services after she impressed her WEP supervisors; or Bobby Lane, an ex-drug felon who found work as a social worker after getting himself off welfare. Through it all, the mayor never lost faith in the teams of city welfare workers. "They clean the parks, they clean the streets, they clean the buses, and they clean the sanitation trucks, and they contribute back to the rest of the society," the mayor told reporters in the early months. "Now what's wrong with that? That is exactly what the social contract is all about."

Once, while marching down Fifth Avenue in a 1996 Labor Day parade, the mayor shouted down protesters who were demanding union protections for WEP workers. "I think this is one of the best things I've done as mayor is to introduce workfare instead of welfare," he said to the crowd. "Work beats dependency anytime," Giuliani declared. In twenty years, he predicted, "you'll thank me for getting you off welfare."

The beauty of WEP was in its potential to alter behavior, if not attitudes. "Work is hard," Giuliani told reporters later. "Getting up every day, making sure you get a job, getting yourself cleaned and ready to do it—there's a certain discipline that's required for everybody in that. And what we were doing in welfare before was training, training, training."

The mayor often relished doing battle with his critics. "I love to watch the reaction of the so-called intellectuals," Giuliani told the *New York Times* in 1995. "The thinking establishment goes into convulsions over the idea that we could ask people on welfare to work, or that we should fingerprint them to prevent fraud. It's almost as if a secular religion had developed in which these are the things you must believe to be considered an educated, intelligent, and moral person."

As his welfare policy took off, the mayor's views on poverty continued to evolve from the systemic to the individual. As a candidate, he had argued that most New Yorkers reluctantly sought temporary relief on the welfare rolls because a sagging economy had left them little choice. As mayor, he rejected explanations based on temporal economics. New Yorkers were on welfare, he said, because they felt trapped in a system that left them with no incentive to do for themselves. A truly responsible government would ensure that families got out from under the burden of welfare as soon as possible. Tackling poverty became less important to him than addressing the very roots of lethargy.

In his later term, the mayor began to speak about the poor as an errant class of citizens, suggesting that some sought food stamps in order to gamble with them, that some women claimed domestic violence in their lives in order to get better housing, and that others might be better off leaving the city for a more job-friendly region. In this way, Giuliani skillfully turned the welfare discussion back to the question posed by the Victorians—who was worthy to receive help, and who was not? The answer, he said, lay unambiguously in work. Those who worked deserved help. Those who did not would suffer the consequences. The public, for the most part, did not disagree.

By the time President Clinton signed the nation's welfare reform bill in 1996, Giuliani had established himself as an aggressive pioneer in the realm of poverty politics. He had altered the public dialogue on welfare. Public aid was no longer considered an entitlement for those caught short, but a privilege for those who worked for it. Benefits were no longer a right, but an impediment to personal independence.

The only time the mayor seemed to diverge from these beliefs was when he spoke about immigrants. He did not share his party's harsh view on immigrants or immigration policy. Loyal to his grandfather's legacy, Giuliani considered newcomers as a group quite apart from the average poor New Yorker. He recognized that immigrants, even the poor ones, were the lifeblood of the city. Without the steady influx of Asians, Russians, South Americans, and Europeans to its shores and neighborhoods, the city's population and economic vitality were likely to stagnate. So when the federal Personal Responsibility Act called for severe cuts in Medicaid and food stamps to legal immigrants, the mayor of New York objected strongly. Giuliani used the sixties lexicon in his appeal, calling these benefits "entitlements" for legal immigrants. He challenged Congress to reinstate them or risk unfairly punishing a vast majority of urban families who were simply trying to establish a foothold in America's economy.

In 1997 Giuliani appointed as welfare commissioner Lilliam Barrios Paoli, a job training expert with a special flair for government administration. The mayor's first welfare commissioner, Marva Hammons, had resigned. A Democrat by practice, a Mexican by birth, and an anthropologist by training, Paoli took over HRA with great reluctance. She disagreed with the mayor on many key points. Giuliani had said on many occasions that the greatest error of past welfare programs was placing training and education before work. Paoli believed that many recipients required English training and job skills classes. She saw the mayor's program as "smack-in-the-face" welfare reform. During her brief tenure, Paoli beefed up child care, brought poverty advocates to the policy table, and mixed language classes and job skills training into the WEP week. "I was naive enough to think I could change things in the beginning," Paoli said from her current office at United Way.

Giuliani recognized quickly enough that Paoli would not enforce his priorities. After a little more than a year, a City Hall aide delivered the bad news to her. "He said I was a great humanitarian," Paoli remembered of the meeting. "He told me I didn't have the stomach for what they were about to do with welfare." A few weeks later Giuliani transferred Paoli to another city agency. The helm of HRA was once again open for a commissioner who shared Giuliani's vision. Emboldened by the rapid changes he had overseen thus far, the mayor wanted to do something more dramatic with welfare, something that would make New York a singular laboratory in advanced social science. Maybe that would mean putting every recipient to work. Maybe, even, that would mean eliminating welfare altogether. That's when the mayor turned to Milwaukee, Wisconsin, to recruit Jason Turner, one of the nation's most conservative and innovative welfare policymakers.

6

Alina

A Cellar Full of Potatoes

few months after Turner arrived, Alina, an elf on a mission, breezed into the Bronx Jewish Community Council in Parkchester. The year was 1998. She was embroiled in yet another Work Experience Program fiasco, not of her own making. Tiny, thin, her round, milky face framed with jet black hair, she looked more like a self-assured sixteen-year-old than a world-weary twenty-three-year-old premed student. Alina took her place purposefully in the social service agency, unfazed by the paint-spattered sink looming next to her head. "Hello, can I help you?" she asked in staccato New York English, graced by the throaty vowels of a Russian accent. In the four years since the Zukinas had journeyed warily up the Van Wyck Expressway toward their new and alien life, only Alina among them had glided into the rush-or-be-rushed pace that suited life on the run in America. She smiled, revealing teeth slightly off-color, a sign of a Russian childhood diet poor in vitamins and minerals. Alina could virtually count on two hands the number of oranges and apples she had eaten as a child. Paradise, she once imagined, meant reading a book and eating a banana at the same time.

This Bronx center was where Alina had clocked in her WEP hours, twenty-two per week for nearly four years. Workfare was never meant to last that long, but Alina had shimmied under the city's radar for all these years doing what was required of her. Hers was a clerical assignment, answering phones and fixing the files. The nonprofit social services office, just two stairs up from the Holland Avenue sidewalk, was once a janitor's apartment. In the public bathroom, a white ceramic bathtub sat prominently on claw feet. Remnants of a shower curtain wobbled overhead. The community center provided everything from meals to furniture to refugee resettlement.

The work fulfilled the mayor's social contract, paying the public back for the $170 a month it gave her in welfare benefits. By WEP standards, it was a relatively good deal. "Could be worse," Alina would often remind herself. One of her Hunter College classmates, a Russian nursing student, worked off her required hours by cleaning toilets in a municipal building for hours every week. The Bronx Jewish center was one of the rare nonprofit WEP sites set up when the city ran out of government spots. Once again, Alina found herself in a "desirable" welfare position, but not altogether sure whether she should feel lucky or used. She worked with many Russian Jews. The office was near her home. Her boss, Sue Tozzi, was gracious enough to let her study between projects. She was able to maintain her straight-A average as a psychology major. Still, for someone so focused on her life's goal, the work was little more than twenty-two hours a week forcefully carved out of her studies. She had tried four years earlier to secure a WEP job in a public hospital. At least then she might learn something she could use. Alina had taken her request all the way to a state hearing, thinking it was reasonable. But she lost. No reason was given.

Now, in July 1998, Alina's welfare case was suddenly submerged in bureaucratic chaos. Her time card for a recent week's work had been lost somewhere between the social services center and the borough workfare office. HRA responded by immediately cutting her off the rolls, no questions asked. Over the last four years she and her boss had seen the welfare rules become more and more unforgiving. "Look how easy it is to knock people off. They don't even bother to find out what happened," said Sue Tozzi, the center's spry director. "There used to be a much more welcoming atmosphere for immigrants. Now there is a factory mentality—just churn them through. There is no sense of these people as people." Alina appealed to the state court for reinstatement. In the confusion, the city tried to move her WEP job ninety minutes away by subway to a Manhattan health office and to increase her hours to thirty-two per week. It would be a disaster for her carefully scripted schedule.

Alina was understandably unhinged. She couldn't manage a forty-hour workweek, including commuting time, and still make it through school. She needed the money and the health care that welfare provided. Yet she had staked her whole existence on the college degree. Suddenly, in this new equation, studying could equal starvation. She was just a few credits shy of graduating. She had no other income. Her father had left the welfare rolls on his own and was supporting the family on slightly more than $10,000 a year delivering meals on wheels for seniors. During off-hours he stocked vegetables at the Left Finger, a mom-and-pop grocery on Lydig Avenue behind the family's apartment. Alina's mother was trying to get a home attendant license so that she could care for elderly shut-ins in the neighborhood. But for now, Lyusya was working for free, cooking kosher meals for a local yeshiva school.

Four years of studying, conquering English, battling her own shyness enough to place in the top 1 percent of the nation's undergraduates, and now this. Without a bachelor's degree, there would be no medical school. But without the $42.50 a week plus Medicaid coverage and a small rent stipend, she would lose her safety net. "I am refusing to show up," Alina said about this new WEP job. "I will take it to the fair hearing." The American welfare system had taught Alina to act out of character—to face down the bureaucrats, to stick up for herself.

As stressed as she was, Alina knew there were others in more precarious circumstances. The steady stream of Russians in and out of the center attested to that. One frequent client, a young mother named Lyudmilla Koritskaya, had come to America with a small child. It was unusual for Russians to emigrate without relatives in America to serve as sponsors at the required $10,000 per person. But Lyudmilla had been alone even in Russia. A friend had once told Lyudmilla, "If you stay in Russia, you die in Russia." So she headed to America, where she hoped, at least, for survival. In her first few months as a refugee, child welfare workers visited her temporary apartment, provided by the New York Association for New Americans, an immigrant aid group. The baby had been crying and crying, and apparently a neighbor complained. Details were sketchy, but city caseworkers took her baby into foster care, and Lyudmilla soon fell into depression, eventually checking herself into a hospital. After twelve months, her daughter was still being raised by non-Russian-speaking foster parents. Lyudmilla was trying to juggle doctors' appointments with welfare fraud detectors and court appointments. She wasn't sure about what had happened to her child, nor about what would happen next.

Another caseworker told of a mildly retarded Russian who was assigned to work a WEP job at Orchard Beach in the Bronx. "Every day a school bus would come to the corner of Fordham Road and pick up all the WEP workers for the day," said Tatyana Karasovitskaya, a Bronx Jewish center social worker. "Every day this man would be the only one to get off the bus to pick up the trash on the beach," she said. "Everyone else, including the supervisor, stayed back on the bus." Knowing no English, the forty-year-old father was evidently more unnerved by staying on the bus as its only white occupant than by working his seven-hour shift completely alone. When his shift was over, the Russian immigrant boarded the bus, and everyone headed back to Fordham Road.

Another woman, in her fifties, was kicked off welfare for failing to show up with her orange vest and broom to sweep trash near the No. 6 subway stop in the North Bronx. Valentina Pogorelova, who was studying to be an accountant, had worked that job for months and suffered heart problems and exhaustion as a result. She was unable to study and sank into depression. A doctor wrote a note explaining her condition, but the welfare center lost it. Her fair hearing was denied. So Valentina and her elderly husband were living off his Supplemental Security Income for the disabled, plus any food vouchers the Jewish center could get for them. "People may leave the welfare, but they still don't have the means to support themselves," said Mayya Ludman, a Bronx Jewish center caseworker who emigrated with her husband and son from Kishinev in 1995. "They find out soon that it's not so easy over here."

The irony was ripe. Russians were probably more accustomed to the perils of government assistance than any other refugees. The Soviet Union was notorious for its inert systems and paralyzing lines. Russians understood about bottleneck bureaucracy. They also knew what real poverty was like, only without the stigma of class and race attached. Wages for all Moldovans were often ten times below the cost of living. Coming from such border-to-border deprivation, Moldovan immigrants felt no shame in expecting a handout in America to get on their feet. They had left everything they knew behind. Some benefits were expected. Why not?

But the new spirit of American welfare reform was difficult for Russian immigrants to grasp. It was all about an underlying "lesson" the system felt compelled to teach—a uniquely American lesson about rugged individualism. Russians saw an irrational program in which men who had been engineers and accountants in Russia were now delivering pizza and doing janitorial work. They had lost not only their income but their social standing. Hardworking women were encouraged to leave their children with strange baby-sitters to learn this WEP lesson about hard work. Students were expected to abandon their studies to sweep streets. Clerical ineptitude seemed to be a deliberate tactic to knock people off the rolls. For the small benefit check, it was hardly worth the effort or humiliation. "It's a circle of stupid things, a complicated mechanism. Reminds me of the Soviet Union," scoffed Yeva Schleyger, one of the center's caseworkers, herself a recent refugee from Moldova. "They should punish people who get addicted to charity," she said, echoing the beliefs of the reform architects. "But why punish someone like Alina?"

A hot August sun beat down on Manhattan's Flatiron District, lighting the unmarked door to the city's Office of Employment Services on East Sixteenth Street. Once a hotbed of labor protests before both wars, then a seedy drug spot during the seventies' fiscal crisis, nearby Union Square Park was now a thriving shopping center for the Wiz, Barnes & Noble, Gateway Computers, and other corporate franchises. Most of the New Yorkers dining on fifteen-dollar tofu dishes in the Zen Palate down the block or sipping six-dollar cappuccinos across Union Square Park in the retro Coffee Shop were unaware of what went on in this behemoth building nearby.

The Office of Employment Services headquarters was where every city welfare recipient went to receive or negotiate a WEP assignment. Like most city offices for the poor, the building tried its best to remain anonymous. Faded black address numbers peeked through inches of grime on the outside of the door. Venetian blinds, broken in several places, hid the goings-on behind the cloudy windows. Only an American flag flying next to the official city flag posted high overhead indicated that this building housed official government services.

The cluster of bargain vendors on the sidewalk outside provided another clue. It could have been a scene out of Alina's hometown of Kishinev. One street peddler spread out his tin jewelry, ties, and sunglasses on a makeshift table, everything for one dollar. Another sold shaved ice doused in neon-colored syrup. A third hawked hot dogs and soda. Idling trucks spewed diesel fumes into the air where dozens of people were milling outside. Some were eating their Italian ices in sundresses and suits, others in saggy jeans and oversized shirts. A steady stream of humanity flowed in and out of OES, navigating around the two dozen garbage bags piled near the door. An old Chevy parked out front with its windows down blasted rap music indiscriminately.

Alina scampered up the concrete steps with a big welcoming smile, right on time. She was dashing from her volunteer job at Sloan Kettering Hospital uptown. For four hours every other week, Alina delivered flowers and answered phones at the famous cancer research center, just to get a taste of an American hospital. She found inspiration in the supplies, the cleanliness, the technical support. In Moldova, patients had to bring their own bedsheets and syringes if they were unfortunate enough to require hospitalization. Doctors expected bribes to provide basic care. Though Alina's duties at Sloan Kettering were strictly nonmedical, they kept her in touch and motivated.

Today, on July 2, 1998, she would hear her fate at an administrative hearing. Would she have to quit school to work the new WEP job? Would she have to quit welfare and figure out another way to eat? Or would the city come up with a reasoned compromise? Alina rushed past the guard at OES, whose shirt was partially untucked. Handwritten signs were taped to the walls, but none gave her the directions she needed. Finally, she found the right line in which to wait for a number. Then another queue to sign in for her appointment.

After about an hour, Alina was beckoned into a small room. A hearing officer spoke into a battered tape recorder in legalese while Alina waited. James Ellerby, Esq., lifted his head to ask Alina her name, then returned to his dictation. "You won your case," Mr. Ellerby said after a few moments. "Do you understand?" Alina nodded. She understood the English, but not necessarily the meaning. His quick dismissal made her nervous. The officer asked her no questions, offered no explanations. Did the city have plans to send her to work on a garbage crew somewhere? Was that what she won? Most hearing officers had been rude to her before. Rudeness she understood. But this one was neither kind nor mean. He was nothing. Instead of tempting fate, Alina decided to flee without answers to her questions and to wait for the next hoop to appear.

Months passed. Alina received word to report to the Parks Department. She sat through the orientation sessions, learning about all the WEP rules—how to show up to work on time, how to properly spear trash in the parks. When it was time to get her assignment, the WEP director of Bronx parks glanced at her sideways, perhaps wondering whether she was old enough to have her own welfare case, and told her to go home and wait for a call. The pixie-like Alina, with her transparent skin, looked as if she could blow away with the raked leaves at the first strong gust. "He took a look at me and said he would find me some job inside," Alina said, laughing. He never did.

Close to Christmas, Alina was summoned down to East Sixteenth Street once again. She was just one or two exams away from the end of the semester. She still wasn't working for her benefits. She would have been working, of course, if her time card hadn't been lost, if the Bronx supervisor had found her a job. But all that seemed well beyond anyone's institutional memory. She now stood personally accused of collecting benefits and not working for them in return. Today she would discover her long-term fate, not just her temporary one. Alina took a seat with four dozen other welfare recipients on the plastic chairs and rehearsed her argument. "Okay, I have experience with the Department of Aging. That's what Sue Tozzi's office is. It's where I can be most useful. I have only a few credits left to graduate. I hear they give some kind of leniency to students who are almost done. They know we'll be off the rolls soon. We'll finally be earning a real living."

After an hour, a robust woman with a Jamaican accent called out an approximation of Alina's name from the corner of the room. Alina followed her to a desk, set among dozens of others in a large room. As Alina explained it, the interview seemed to turn sour before she even sat down.

"So why aren't you working?" the caseworker asked Alina as she riffled through some papers.

"I'm waiting for an assignment. I've been working for the Department of Aging, and then they bumped me off. This is the fifth time I've been here. No one can find a place for me," Alina answered, in her crispest English.

"I will ask you again, why aren't you working? What are your future plans?"

"I will go to medical school," Alina said meekly.

"I didn't ask you about school, I am talking about work. What are your plans for your life?" the caseworker retorted, her melodious voice rising in anger.

"I told you, my future plans are, I am going to medical school. I will become a doctor," Alina answered, trying to hold her own.

The caseworker fumed and sent Alina back to the waiting area. Two hours later she summoned her back. Still angry, the caseworker lectured Alina on the work ethic and the art of being a successful physician. "You need a serious attitude adjustment," she said, enunciating the t's in *at-ti-t*ude and adjus*t*-men*t*. "You need to give in order to receive. If you want to be a doctor, you have to be more polite."

Alina responded that she was usually polite, but that she did not like to be treated with such rudeness. A long pause followed. Alina couldn't believe her own brazen retort. She began calculating in her head how she was going to finish school without her benefits when this woman dribbled her out the door. She was even more worried about losing Medicaid. There was no way she could afford doctors' visits on her own.

The caseworker let out a muted whistle and adjusted her arms on the desk. "I don't have to take anyone's attitude. You were bumped out of the Department of Aging? Well, that's exactly where you're going back to now." She huffed, as if delivering Alina to her worst nightmare.

Alina tried not to show her relief as she scuttled out of the waiting area and down the elevator as quickly as she came. This meeting had bought her some more time. She couldn't wait to put all these annoying caseworker encounters behind her. Some workers were pleasant, like the young black woman at the Willis Avenue Income Support Center in the Bronx who told her quietly never to give up her schoolwork, to stick with it as long as the system allowed her. There was another Russian caseworker at the Sixteenth Street office who really tried to help. But most of them seemed overworked and uninspired, bent on putting in their hours, following orders, making the unspoken quotas.

Her face drawn from the stress of this latest exchange, Alina headed for home by subway. Every encounter in English, particularly the hostile ones, left her feeling drained. It was hard work for her. Before speaking, she was still translating everything from Russian to English in her head. Even though she thought her courses were easier in the United States than in Moldova, she had to spend twice as long studying here because of the language difference. Writing a paper took ages. There wasn't time to rest. She had several research papers due and MCATs to study for in order to have a shot at medical school.

And her welfare clock was about to expire. In one year she would be thrown off welfare for life.

A lina's mother was waiting back at the apartment in her housedress and sandals. Thin pancakes and jam were ready on the kitchen table. Lyusya was restless in her sixth-floor Bronx apartment. Offering homemade fruit juice along with opinions on the sins of American fast food, she darted around her kitchen practicing the magic of creating meals from scratch with meager ingredients. She turned cucumbers into pickles, sour cream into farmer cheese, cheap chicken parts into soup—a Moldovan Martha Stewart on a welfare budget. It pained her that both her daughter and son were so thin. "Too skinny, my children. They don't eat enough."

A handsome woman, plump, with short graying hair and big brown eyes, Alina's mom was an energetic force, used to earning her own living, not being dependent on her husband or her daughter for food and shelter. Alina looked even more like a waif in her mom's solid shadow. Lyusya seemed older than her forty-four years. Perhaps that was because Moldovans aged faster. Retirement age was sixty back home. The average life span for men was sixty-two. "My only regret, I need a job," said Lyusya, smoothing her dress. "I love to work, to make food, with love. That's what I did in Moldova. For twenty years I worked in the same place. I was a supervisor, the only Jewish woman. I used to be independent from my husband."

Just then, her twelve-year-old son, Moshe, bounded through the door, all arms and legs, home from school. The pale boy with cropped blond hair headed straight for the donated computer in the corner, gluing his ice-blue eyes to the screen, his new obsession. Last week's obsessions were taped to the wall over his work table: a poster of Rabbi Menachem Schneerson, the late Lubavitcher spiritual leader from Brooklyn, plus a magazine photo of the Spice Girls. Though he struggled in school socially and academically, Moshe had embraced the American experience more fiercely than anyone else in his family. His was a blend of new religious freedom (orthodox for now) with technology and kitschy pop stars. Lyusya felt compelled to stay home in order to help Moshe adjust. "I have to be close by for Moshe. I can't take a job just anywhere," said his mother, within his earshot. She worried about his grades, his behavior, his general well-being. "He doesn't want to make his homework. He doesn't want to learn. I have to be nearby to make sure he does these things. He is a bad boy." Moshe stared even more intently at his computer game.

Eventually diagnosed with a mild learning disability, Moshe was held back two years in school. At age twelve, he was in the fourth grade, more content in the smaller yeshiva than in the public school. A scholarship helped cover the Jewish school tuition. Lyusya volunteered in the school kitchen to offset the rest of the expenses. Moshe was doing well enough now that Lyusya's role was beginning to diminish.

Immigration had turned Lyusya's life upside down. She sat on the couch in her living room, detailing her persecution with gusto. The room was pleasant enough, comfortable, with a television at one end, the computer at the other. Lace curtains hung on the windows that overlooked Pelham Parkway. "Look, this is garbage, this furniture, not like I had back home," she said, motioning around the room. Two cloth-covered couches, apparently from someone's discard pile, lined both walls. One of them served as Moshe's bed. There was another bedroom for the parents. Alina slept in a converted pantry off the kitchen.

Alina's two-year engagement to Mark Filizov, a young man she originally met in Kishinev, made her mother ashamed. "In my country, if you wait two years to get married, something is wrong. It's humiliating, engaged for so long." Even her daughter's medical ambition sparked anxiety instead of pride. "The most important thing for a woman is to be a good homemaker," Lyusya declared. "It's too hard if you are in medical school. It's too hard if you are a doctor." Alina smiled in silence from across the room, apparently unfazed by her mother's disapproval.

"My mother used to have the power in the family. My father would defer to her on all decisions," Alina said later, with a twinkle. At twenty-three, Alina was experiencing a delayed adolescent breakup between mother and daughter. "Now she tells me I am the boss of the family. She blames America, of course. Whatever she says, I do the opposite. She says, 'Have some sour cream on your borscht,' and I say no, even if I was thinking about having some. Even my father doesn't listen to her. He does what he wants." Alina's mother wanted her nearby, interpreting TV for her, filling out applications. "I have to translate everything for her. I get tired of telling her what's going on, where to fill in her name. It's not relaxing for me."

Lyusya scoured through drawers looking for photos of Kishinev to show the markets, the schools, the streets, and the apartment buildings that had shaped their lives. Without such pictures, Kishinev was difficult to visualize. The city's history went back to the fifteenth century, when it was first settled as an obscure provincial town. Its parks and social life were revealed in the 1820s when Alexander Pushkin lived there in what he called the "deserts of Bessarabia" as a poet in exile. Some Kishinev scenes were described in his works *Eugene Onegin* and *Gabriliade*. The former Bessarabian town figured in historical texts when it bounced between Romanian, Turkish, and Russian control. But modern Moldova—a country carved from the spoils of the last world war—was more or less a mystery to the West.

Lyusya pulled out a handful of photos, most of them from the town of her childhood, Sharogrod, Ukraine. One sepia-aged picture of a one-story stone house caused her to quicken her breath, for a moment. Lyusya was a young girl in the photo, standing on a log in a loose cloth dress and boots with her father, mother, and siblings. This was her

Ukrainian home. "That is my father," she said, pointing to the stiff man with a beard. "I feel very sorry. He worked so hard his whole life. Once, I told my father, 'You work hard, but we don't have much things,' " Lyusya remembered. "He was angry and said, 'What are you talking about?' He took me to the cellar under this house and showed me all the potatoes. 'Look how much potatoes we have. You have clothes. You have school. It's enough.' " Her father died soon after in a Ukrainian labor camp, shot by the Nazi occupiers.

In order to see and smell Moldova like that, to get a renewed sense of a country where prosperity meant a cellar full of potatoes, Alina would have to get back on that dreaded airplane for a visit.

7

Brenda

Urban Horizons

A few parceled neighborhoods west of the Zukinas', Brenda Fields began, at last, to arrange family photos on the walls of her new apartment. She pulled out old bedspreads, forgotten artwork, and familiar dishes from the collection of mashed boxes in the hallway. It was a comforting task, one she'd put off for too many weeks. Brenda's fear of losing Ty in a custody battle to his father had kept her off balance from the day she moved into this apartment. When she wasn't chasing down child care and welfare appointments, she was going to court to address legal questions. Ty was a bright child, eager to learn his colors, his numbers, his days of the week. He marched through his Bronx world of hallways and sidewalks with a precocious sense of himself, striking up "whacha-doin' " conversations with adults, asking how the Xerox machine made copies, how the train blew its whistle. Brenda didn't want an ugly custody battle to clutter this child's path. The day of the hearing, Brenda arrived at Bronx Family Court before the guards opened the doors. Four hours later, she left for home. Ty's father did not show up. The case was dismissed, at least for the foreseeable future.

Brenda could finally concentrate on her newfound residential luck—a place in this renovated apartment building that had been named, rather hopefully, Urban Horizons. Anyone could spot its classical beauty from Babe Ruth Plaza and from the curvaceous stadium Babe built, metaphorically, along the Harlem River. Urban Horizons towered like a Florentine monument over the stray dogs and chop shops on Jerome Avenue, over the bodegas and supermarkets on East 167th Street, and over the hole-in-the-wall weed shops on Walton Avenue. Brenda had a strange affinity with its phoenixlike past. Now she banked her own family's future on the building's neo-renaissance present. "I may not go to church much," she said, taking a break from hollering at Ty to calm himself and unpacking pots and pans, "but I light a candle every day for giving me this next chance." It was the fall of 1997.

The story of the former Morrisania Hospital was the story of a South Bronx that no one ever imagined. It was built in the twenties, "the golden age" of the borough, when the nation's best architects strove to construct monuments to herald new ideas and ensure their own immortality. The city asked Charles Meyers, most noted for designing the Tombs and Manhattan's House of Detention for Men, to create a municipal hospital campus on a full city block just west of the palatial art deco apartments on the Grand Concourse. The neighborhood was mostly working-class Irish in those days, with a smattering of Scandinavians on the bluffs. Well-to-do Jews—those who had long ago worked themselves out of the "lung blocks" on the Lower East Side—lived at the crest of 168th Street along the Grand Concourse. Blacks began migrating into the area south of Crotona Park called Morrisania, the former estate of Lewis Morris, a signer of the Declaration of Independence. Children of each ethnicity played potzie on the sidewalks in front of the busy hospital, or punchball with their spaldeens in the relatively safe streets. Built of concrete and steel, dressed in blended blond bricks, topped with ruffled Spanish roofing tiles, Morrisania Hospital was meant to last.

The Great Depression cast the first pall over the neighborhood's idyllic stability. Then, in the fifties, parks commissioner Robert Moses built a mile-long highway, slicing a concrete path straight across the Bronx from the Hudson to the East River. The construction of the Cross Bronx Expressway, otherwise known as "Heartbreak Highway," cast five thousand families out of their South Bronx homes and jack-hammered entire neighborhoods into near-extinction. Middle-class families migrated north, while hundreds of thousands of poor blacks and Puerto Ricans displaced by Manhattan slum-clearing projects moved into the disheveled apartments left behind. Banks left, businesses fled, battered buildings were torched instead of fixed, landlords cashed in on insurance claims, and scavengers stripped the scorched hulks for scrap metal.

The South Bronx soon became a national symbol for urban decay. The once-grand borough nearly burned to the ground. And the once-grand Morrisania Hospital remained as a hulking witness to its decimation.

Along with neighborhood firehouses and schools, the city abandoned the hospital in 1976, placing cinder blocks in the doorways and ~~b~~ards on the windows, and wrapping its perimeter in a chain-link ~~fenc~~e. Over the next two decades the photojournalist Camilo Jose Ver~~a~~ visited the potpourri of scavengers and squatters who hacked ~~throug~~h the barriers to inhabit the ruin. Some mined the wreckage for ~~copper a~~nd bronze wire. Some lived in its blood donor and autopsy ~~rooms al~~ongside the test tubes and the rats. Still others used the place ~~as a fligh~~t base to sell crack or prey on passersby. Through it all, ~~Morrisania~~ Hospital's sturdy beginnings helped it defy the fate of ~~less-solid~~ structures. It did not crumble. It did not burn down. It ~~was not replace~~d by a parking lot or an auto repair shop. Instead, as

Bronx Borough President Fernando Ferrer would later say, the ghostly shell would turn from "an open sore into a new urban horizon."

Nancy Biberman used to eye the open sore every day on her way to work in the mid-eighties. The building was hideous, no question. But to her eye it was also filled with renewing possibilities. A descendent of Ukrainian Jewish immigrants, Nancy grew up with an unwavering sense that it was up to her to change the world. In 1968 she had emerged as a leader of the infamous student strike at Columbia University (that is, if leader meant "girlfriend of leader," she hastened to add), one of the nation's most volatile anti–Vietnam War revolts. She moved on from those heady days to forge a career for herself as a poverty lawyer in Manhattan. She helped welfare recipients get their entitled benefits in the seventies, she protected elderly residents from being evicted by co-op developers in the gentrifying eighties. Her daily tasks seemed to always place her in the path of a booming economy vaulting over those who could not afford to live within it.

In the late eighties the city was flush with money, pockmarked with half-empty buildings, and teeming with homeless people on welfare. By the millennium, more than 60,000 new apartments would be available in the Bronx, and more than 600 businesses would be created, mostly by nonprofits with public funds. The majority of the city's homeless would eventually flock to the South Bronx to find a home. Nancy was part of that Bronx rebirth, working to develop 23 buildings into 722 apartments in the burned-out Highbridge section of the Bronx, on the cliffs east of Morrisania Hospital. It was an enormous low-income joint project between the Housing Preservation and Development Agency and the Archdiocese of New York, one of many renaissance efforts for the deserted area.

Nancy spent time touring the neighborhood with a friend who had once lived there when the community was thriving, the streets packed with children looking for the next stickball game. "It was painful for her to remember, 'Here was the school, here was the park, here was

the pizza place where we all hung out,' " Nancy said. "There was literally nothing there. It was gone." Her friend's childhood memories had
been ground into gravel. When a contractor pointed to an empty lot
saying he had built a new housing unit there just a decade earlier,
Nancy realized that people needed more than brick and mortar
boxes. They needed reasons to forge ties, reasons to stay put. The former legal aid lawyer made a modest effort with the Highbridge housing project to fill that gap. The project created day care, social
services, after-school programs, a health clinic. But when its Catholic
partner insisted that no AIDS or abortion information could be distributed inside the health care clinic, she quit. "Knowing what we
know about AIDS rates and child birth rates in that neighborhood,
I just couldn't stay," Nancy said. "I mean, could I look a pregnant
seventeen-year-old girl with two kids in the eye and say there is nothing we could do for her?"

So Nancy turned her attention back to her fantasy ruin on East
168th Street. This could be a place to start from scratch, to invent
housing the way it should be done. She convinced an architect and
a contractor to come take a look with her. They cut a hole in the
chain-link fence and crawled through. Inside, the unlikely group
waded through waist-high debris and animal carcasses, fending off
dogs and wary squatters, to find that the beams and floors were
miraculously unscathed. The building had to be gutted, yes, but not
razed. She gathered Barbara Petro-Budacovich, a nurse with a Ph.D.
in social services, and Ismene Speliotis, a financial planner, and
the three hashed out a vision. They would form a nonprofit corporation: Women's Housing and Economic Development Corporation
(WHEDCO). Together they would drum up the $23 million needed
to fund the project.

The wish list was easy to compose. The hospital could be converted into 132 apartments. The old ambulance garage could become
a state-of-the-art fitness center. The old hospital kitchen and bakery

could be transformed into a day care center. The ambulance entrance could house a commercial kitchen. The bulk storage area could become a health care clinic, counseling center, and job placement office. Not just job placement, but job creation. Counselors could train women for day care jobs, catering, small-business ownership. They could identify opportunities in hotel operations, computers, and restaurant services. These would be real jobs, with a future attached.

Several years before workfare became the city's and the nation's buzzword, Nancy and her coworkers were designing their own work programs for welfare recipients. Work, they strongly believed, was central to poor women's futures. They focused on women because the vast majority of this neighborhood's welfare homes were run by single women. The name of their corporation was, after all, Women's Housing and *Economic* Development. For Nancy, the most positive part about welfare reform was an acknowledgment that recipients could earn extra money on top of receiving their small benefit check. Before the change, people on public assistance were essentially punished by being cut off if they earned extra money baby-sitting or delivering groceries. The poor considered the extra work a necessary lifeline to make ends meet. Welfare officials considered it fraud and took steps to cut off the offenders' total package—no more rent help, medical insurance, or food stamps. So the extra job experience was rarely worth the small improvement but larger risk of deprivation it caused. The new 1996 rules encouraged people to seek jobs. Welfare would keep the benefits coming (at reduced rates, for a brief period) as long as the recipient could demonstrate that he or she was actively searching for work.

In her fourteen years as a legal services attorney, Nancy had never met a welfare recipient who wanted to stay on welfare. Everyone wanted to work. But work meant something different to Nancy and Brenda, and Christina, and Alina, than it did to city officials. It didn't mean taking any minimum-wage job with few benefits. Work meant financial independence for women in jobs that promised to be spring-boards out of poverty.

WHEDCO's philosophy was to find out what the community wanted and then provide it. Years later, under WEP, the city's philosophy would be to inform welfare recipients what they needed and then force compliance. The city's assumption was that welfare recipients were on an equal playing field with any American in search of self-sufficiency. WHEDCO believed these women needed far more assistance. The lives of the poor tended to be more stressed, their family networks more fractured, their literacy rates far lower, their job opportunities fewer. Morrisania's unemployment rate was 17 percent in 1997, nearly twice the city's average, and four times greater than the nation's. Close to 80 percent of its children were born into poverty. One-third of the neighborhood's families depended on public assistance as their sole source of income. Children's asthma, pneumonia, and infant mortality rates in the area topped the city's charts. For its eight thousand preschool children, the neighborhood could provide only twenty-two hundred day care spots. "As mothers, we knew what it's like to have to juggle work with a sick child," said Nancy. "But most of us can afford a string of baby-sitters, or quality day care. Most of us have family that can pitch in financially if we have an emergency. These women don't always have that kind of backup."

Miraculously, nearly every wish on WHEDCO's list evolved into reality, despite the forces of race, gender, and politics working against such success. The first hurdle was to shore up political support in the treacherous trenches of the Bronx. "We had a downtown white problem," in Nancy's words. "We were three white girls from Manhattan" running a new, untried corporation. The local community board told this group of pale neophytes that the Bronx had enough housing. It really needed more upscale businesses, finer restaurants, and more schools to relieve serious overcrowding. Another rival plan surfaced with backing from a local Hispanic politician, who thundered about the women from nowhere moving into the Bronx.

Then providence interceded. In her search for allies, Nancy ran across a group of parents who had been seeking a permanent home for their local bilingual school ever since it was launched in a supermarket some fifteen years earlier. The parents had their eye on the old Morrisania Hospital. But they couldn't see beyond that ten-story carcass. What would they do with it? They couldn't tear it down, since it had historical status. They couldn't convert it into a school. It was an awkward size. WHEDCO could solve that problem. It could transform the white elephant into a home, while the nurses' quarters on the southern end could be torn down to make way for the school. The city block would become a beautiful educational park, with a unified theme and unified buildings.

Now that Hispanic parents were signed on, the plan suddenly carried more panache for the Bronx-born-and-bred borough president, Fernando Ferrer. After fending off aggressive lobbying by his Hispanic colleagues, Ferrer finally agreed to throw his support behind Nancy's ideas. The community board followed suit. The city agreed to turn the site over to WHEDCO in 1994 for one dollar. At the same time the city pled its own brand of poverty—there was no money in the budget to help fund its reconstruction. Nancy turned to the state instead for financing. Part of her plan was to reserve one-third of the apartments for the formerly homeless. That was enough to convince the state's Department of Homeless Housing Assistance to pledge a generous $4.5 million. The next pot of money WHEDCO went after was the State housing agency's contribution to New York City. Just a few months before the gubernatorial election, Governor Mario Cuomo committed over $6.5 million in housing money, the single largest donation to one project in the agency's history. The rest was raised by private equity investments through a low-income housing credit program.

But the celebration was short-lived. In November the top offices in the state and the city were lost to Republicans. Many projects approved by Cuomo were canceled by his successor, George Pataki. "We thought we were doomed," Nancy said from her office perch on the tenth floor, a bright terrace once used as a rabbit laboratory for pregnancy tests. "Then Brent Staples wrote a beautiful story about us in the *New York Times.* The day after it came out, the governor's office called." Two years later, in 1995, Urban Horizons broke ground. "We got in right under the Republican wire." Not a penny of city money was spent to develop city-owned land.

Using mostly Bronx construction workers, Nancy gutted and rehabbed at cost. With historically faithful results, in sixteen months' time, she beat her own self-imposed deadline. Nancy was defiant about the dignity of details. "I wanted it to be beautiful," she said. "The state was happy with ugly as long as it was cheap." Nancy won the battle for an elegant lobby with glass light fixtures. Terrazzo floors were poured with concrete, then dried, then flecked with chips, then smoothed out, then lined with metal strips. "It took forever, but it will also last that long." She argued for chair rails and sconce lighting in the hallways, to offer a sense of "home" instead of "institution." She convinced the state to install solid oak kitchen cabinets with tile back splashes instead of the standard-fare particle board. Some apartments featured kitchen islands. "It's the same money in the long run," she said. "This way, it's gorgeous and inspiring."

Nancy didn't win every fight with the state. In one inspection the state inspector concluded that the apartment ceilings were far too high. Nancy tried to counter with a practical approach, saying this used to be a hospital with twelve-foot-high ceilings that could not be altered. No go, said the state inspector. The ceilings would have to be dropped to fit the low-income housing standard. "I couldn't believe what I was hearing," she said. "The poor did not deserve these magnificent ceilings? And we would have to spend more money just to make sure they don't get them? It spoke volumes about their attitude toward these families."

Exactly two decades after the city sealed Charles Meyers's neo-renaissance masterpiece in chain-link fencing, Nancy Biberman and Barbara Petro-Budacovich unwrapped it once more in 1996 for the community to use. The building was in vintage 1920s shape, barely scarred from its recent odyssey into hell. The only feature deliberately missing from the facade's carefully restored terra-cotta detailing were the words "Morrisania Hospital." Instead, underneath the plaster medical emblem of a snake coiled around Asclepius's walking staff, was a simple address: 50 East 168th Street. The WHEDCO staff had planned for years for this opening day. Tenants were selected by lottery, by poverty-level income caps, and by interviews from a pool of three thousand applicants. Like the neighborhood, more than one-third would be on public assistance. Homeless families, like Brenda's, were interviewed in their shelters by social workers to determine their needs and to weed out those with long-standing substance abuse problems. Only eight families moved in that first day, part of WHEDCO's careful plans to ensure a smooth beginning.

"I'll never forget hearing my name screamed in the hallway that first day," said Barbara. "There was real fear in that scream. I ran to it, thinking, 'What on earth? It's only the first day!' " A small child was lying face up in the hallway, turning blue. The mother was speechless with fear beside him. She was no more than a teenager herself and didn't speak a word of English. As a trained nurse, Barbara recognized the situation as a cardiac seizure. She slipped her hand into the child's mouth to start resuscitation. After a few professional breaths and chest poundings, the child revived. "I thought, 'Wow. What if I hadn't been there?' Welcome to the Bronx. We planned up the kazoo, but we can't plan for everything."

Beneath the residence, reached through the old ambulance entrance on the first floor, lay the real heart of the three women's plans—a labyrinth of people and programs pumping vital resources to the residents above and the community beyond. It was an economics laboratory, reflecting a series of experiments, with the implicit admission that problems were unexpected and not easily solved. There was no way, as Barbara had said, to "plan for everything." The lab was designed with the same attention to dignity that went into the cornices and window arches of the residence above. The idea was to revitalize the neighborhood surrounding Urban Horizons by incubating new jobs and providing quality child care for its working poor. Family and mental health care would be available for whoever needed it. Employment agencies would train clients in the art of interviewing. Computer classes would lift students into the age of the Internet.

Budding entrepreneurs could find help percolating their own day care or catering businesses here. Still others could just drop by for lunch. Those who needed more structure in their day could visit for a job match service or an internship. The separate entrance was important, Nancy said, so that residents upstairs didn't feel they lived in a program. By the end of the millennium, more than three thousand Bronxites had filtered in and out of this former eyesore, making it a beautiful and organic part of the neighborhood once again.

During this period New York State sat on more than $1 billion in unspent federal welfare money—one of the biggest pots of untouched dollars in the country. Nationally, a few states were beginning to spend their surpluses on programs that helped move the poor into work. Five states allowed new workers to keep their welfare benefits for at least a year, to pad their small paychecks. But the vast majority of states were hoarding up to $7 billion total in anti-poverty money. And New York was one of the worst offenders. Some lawmakers were hoping to use the funds as a safety cushion for those still left on the rolls once the federal deadlines passed. Others were pushing for general tax rebates. The battle raged in Albany between Republicans, who wanted to use much of the money to plug budget holes, and Democrats, who wished to use it for literacy classes, computer training, and day care. Nancy and other nonprofit directors had a thousand ideas for such a windfall. She figured she could serve three times the number of people with the extra money. It angered her that the state was sitting on funds that might help the poor find self-sustaining jobs. But the deadline clock for welfare recipients was still ticking. Families would be cut off once their five years ran their course, whether or not that money was spent.

Brenda could not afford to wait to see what would happen next. WEP would catch up with her in a matter of weeks, cutting her job search time in half every week. She knew that her own welfare clock was running out. She needed decent child care for the rambunctious Ty. Every week she checked to see whether WHEDCO had managed to open the federally funded Head Start preschool in the basement. Every week she was told it was coming, soon. In the meantime she decided to cast her lot with WHEDCO's employment service, America Works. It wasn't her first choice. She preferred to go to community college, to build her skills in computers or train as a medical assistant. Without such education, her choices were limited. Brenda had worked in several catering halls over the years. She couldn't list exotic dancing on her résumé, but she did put in legitimate time as a cashier at two department stores. Years before, when her daughter was a little girl, Brenda went through the America Works office in Manhattan to acquire a food service job. It wasn't her favorite experience. But she was willing to give it another chance.

If nothing else, she could call a magnificent building her home. Morrisania Hospital could lift the spirits of those still struggling outside its sanctuary.

8

Brenda

Trabajos/Jobs

renda raced around the rail yard district south of Yankee Stadium searching for the welfare office one snowy November day. Ty looked like a miniature stack of tires in his black puffy coat, his little work boots a blur as he tried to keep pace. "Come on, Ty. We're late," his mom said as they passed the Western Beef warehouse for the second time. Finally, she noticed a beat-up orange truck selling sodas for $1, cheeseburgers for $1.75. It was parked in front of an unmarked door on Rider Avenue. "There it is. Come on." Brenda was late for her appointment with BEGIN—the city's name for its decades-old welfare-to-work program. Before welfare reform, a fairly small percentage of able-bodied welfare recipients whose children were six and older came to BEGIN offices—an acronym for Begin Employment Gain Independence Now—to get workfare assignments and look for permanent jobs. Now BEGIN workers handled work assignments for nearly every recipient with newborn infants as young as twelve weeks old. A handwritten sign next to the elevator urged people to leave their children at home.

A month before Brenda had moved into Urban Horizons with a list of priorities. First, she'd get her paperwork in order. Then she'd bring Ty to day care. Then she'd find herself a good job, no matter what. No one had to convince her that welfare was a demeaning choice when financial independence was an option. She believed in the new welfare reform message—that work was the answer. She wanted nothing less than to head off to a decent job every day, knowing she was providing enough for Loreal to go to college and for Ty to stay up to size with his shoes. She didn't want to depend on anyone else to supply a generous stack of presents under her Christmas tree. But here it was November 1997, and she was still 0-for-3. All the day care centers around her were full. She couldn't look for work with Ty in tow. She was still trying to catch up with welfare meetings. Her blood pressure was spiking from all the tension. She could barely make it on $290 a month in cash benefits plus $207 in food stamps, and soon these might be taken away. She had four years left on her welfare clock. But that didn't bother her. Her more immediate deadline was to meet WEP standards so that she could maintain her benefits and independence. If she didn't work WEP, the city would cut her off. If she did work WEP, she would have no time to look for a real job. Brenda had about thirty more days to come up with a job or find child care, or both.

Brenda had already missed two WEP appointments: once because the welfare fraud bureau was paying a required home visit, the other because Ty had a bad case of conjunctivitis. One more screwup and her check could be docked. She riffled through her Avon bag full of documents and waited her turn. Her eyes traveled to the two oddly incongruous decorations on the wall. A travel poster called "The Ville de Nice" advertised an idyllic French Riviera landscape, a champagne and foie gras scene that the crowd on Rider Avenue would probably never experience. The date on the poster was 1969. The other poster had a more recognizable image. A handwritten sign taped above a metal tray with twenty slots read: "Trabajos/Jobs." All twenty were empty.

Ty scanned the room for a possible playmate, then pulled out a Hot Wheels truck and a pad of paper from his mom's bag, settling down on his knees for the duration, using the chair as a table. "Mom! Mom! I'm doing my homework. I'm drawing your blood. I'm drawing your veins." He yanked on his mom's sleeve, begging her to look at his careful scribbles. This was a boy who was desperate for nursery school. His cheerful patter pierced the dullness that had long before glazed over the waiting crowd. Next to Ty, a woman was tweezing her eyebrows. Another was painting her nose with something that looked like clear fingernail polish while reading a mystery novel. The floor was littered with crushed Oreos and potato chips. A young woman in the first row carried on a loud conversation with anyone who would listen: "Some gets their check and spends it on twenty-dollar tattoos and fancy nails. Not me. No, ma'am. I spend it on my kids. Tell that computer I don't have no baby-sitter. I'm not leaving my kid with no person who kills kids." Some listeners nodded. Others rolled their eyes.

"This place is mad slow, mad slow. Give me *your* job," the woman told a passing welfare worker dressed in a bright yellow African dress. Several people snickered. "You people need some help."

Finally, a tall, round-faced man called out "Brenda Fields" from the end of the room. As he led Brenda to his desk, he tried to make chitchat with another visitor. "What's your name? Are you an American?" For some reason these questions set off Brenda's temper. She detected an attitude, whether he meant it or not. "What's your name?" she asked the caseworker, clearly annoyed.

"Charles," he answered.

"Charles. And what's your last name?"

"Charles Charles," said the Haitian immigrant, with a big grin.

That did it. Brenda snapped. "Is that a joke? Because if it is, I want to tell you this is serious business for me," she said. "This is my livelihood. And I do not appreciate you making jokes."

Charles shrugged off this clash of cultures and flipped rapidly through Brenda's files. She tried to explain that she'd worked WEP in the homeless shelter. That she was going into a job search program. That she had tried several child care centers and they were full. But Charles had already stopped listening. He said it looked as if she had already been in and out of too many job programs. He told Brenda she had two weeks to find child care and handed her a list of a dozen child care providers and phone numbers. What he didn't mention was the agency pressures forcing him to move as many recipients into work activities as he could per month. HRA gave him little more than lists of day care centers, most of which were either full or too far away.

"Oh, yeah. He wants me to leave my child with any old person," Brenda muttered on her way out. "I'm not gonna do that. What kind of mother would I be if I did that?"

Adequate jobs and health insurance for the working poor were important. But when it came down to it, decent child care was the stickiest issue facing policymakers who expected to end welfare as they knew it. The hope was that states would recognize their ragtag child care services for what they were and use surplus welfare money to replace them with something substantial so that poor mothers could go to work and their children's lives would be enriched in their absence. But only a handful of states, such as Wisconsin and Minnesota, were spending their federal dollars on expanded, quality child care. There was growing evidence that poor children were generally worse off than before.

The United States entered into this revolutionary welfare experiment at a distinct disadvantage. It was known for having one of the least extensive, least adequate child care systems in the industrialized world. One national study in 1997 found that nine out of ten American

day care centers did not have basic toys, books, safe hygiene, or adequate numbers of adults per child. Forty percent were actually haz-
ardous, not only to children's health and safety but to their emotional
and intellectual development. Three years later a team of researchers
from Yale and the University of California at Berkeley found that one
million more poor toddlers were enrolled in care as a direct result of
welfare reform. Most of that care was of such low quality that the children were already showing signs of developmental delays in centers,
and even more pronounced problems in the homes of state-paid baby-
sitters. Many toddlers in the study of one thousand homes spent their
hours wandering aimlessly or watching television. Given a choice of
three pictured objects, fewer than two toddlers out of five (the national
norm was four out of five) could identify the picture of a book.

In New York City the strain on kids was evident. Before the city
began calling welfare mothers to work WEP jobs in 1994, it already
counted some twelve thousand poor children on its waiting list for subsidized day care. Five years later that number had more than doubled.
In 1997 the state finally unleashed $64 million from its $1.4 billion in
welfare surplus to expand child care. However, despite the new money,
only thirteen thousand new spaces were created, one-third of the anticipated need, according to Citizens Committee for Children, a nonprofit research group. One solution was to encourage welfare mothers
to leave their children with baby-sitters instead of the pricier centers.
About 80 percent of mothers working WEP were taking that advice.
Calling it "informal" care, HRA paid caregivers $75 per week (as
opposed to $215 a week for center care) per child. Advocates called
this "unregulated" care and tried unsuccessfully to pressure officials to
provide monitoring. In the meantime some nonprofits like WHEDCO
began training some of these "informal" baby-sitters in the art of operating a safe and enriching environment for kids. Even with this cheaper
baby-sitting network, the crunch for care was so severe that it
remained the most cited barrier for moving welfare parents into work.

After Charles's warning, Brenda took one more trip to the base-ment of Urban Horizons just to see whether her best option—the promised Head Start nursery program—was any closer to opening. That's when she noticed an ad on the bulletin board for "Jesus Loves the Little Children," a family care center in the home of a South Bronx woman. Welfare would pay fifty dollars a week for Ty to go part-time to Olga Baez's house while Brenda attended the America Works job program for a month. That would buy her four weeks' exemption from WEP. Her hope was that in one month's time she would have herself a job.

America Works was a job program Brenda knew. She was familiar with its strict rules and professional-like routines. And it couldn't be more convenient, right under Urban Horizons. America Works was an unusual addition to the welfare-to-work stew at WHEDCO—the only for-profit company in the mix. Begun in 1984 by the anti-poverty activist Peter Cove and his wife, Lee Bowes, a social worker, the $7 million agency worked as an old-style head-hunter operation for low-income workers. It trained workers, then matched them, for a fee, with company jobs. An early believer in the work-first model, America Works was once a lone player in the business world. Welfare reform brought hundreds more upstart companies into the market, each vying for a share of the state's hefty federal welfare-to-work dollars. America Works was gradually dwarfed in the stampede. In the mayor's last term, New York City attempted to draft contracts worth half a bil-lion dollars with a number of other firms pledging to find jobs for wel-fare recipients.

New deadlines locked America Works into a race for jobs with sub-sistence wages. Before welfare reform, the city gave America Works four months to find jobs for its clients before cutting their benefits. After the new rules kicked in, the deadline was four weeks, about two months less than the average time it had required to match a welfare recipient to a job. There was no longer any time to improve clients' chances in the work world by teaching new skills in computers or medical care. (After Brenda left, the city clamped down even harder, requiring each America Works participant to work two days a week in a WEP job on top of the job search and training.) The accelerated schedule allowed only enough time for welfare clients to get out and interview for jobs. It did not factor in time for women to complete either a high school or college degree. A high school degree would translate into nearly 30 percent higher wages, according to labor statistics. A college degree almost always guaranteed a career path out of poverty. Like the welfare reformers of the nineties, Cove and Bowes believed in work first, education and training second. Work was the key to ending poverty. Sales reps, as the staff called themselves, only had time to polish up the women's images, teaching them to dress, to speak up, to feel good, to compose a résumé, to shake hands, to embrace the urgency to work. Considering that the average client came to them reading at a sixth- to eighth-grade level, America Works did not always get the job match done.

Teacher Didria Brown glided into her motivational class about ten minutes late one summer morning in 1998. Groomed in a tailored pants suit, her gray-flecked hair flat against her head like a stylish helmet, Didria put on a face of mock misery. "Go ahead, ladies, say it. Bad teacher. Bad teacher," she said. "I'm late. I tell you never to be late. But I can't stop people from getting sick on the train." Brenda was waiting in the class with seven other women, each dressed in their job-interview best. They all chuckled. Didria was a favorite.

"Okay, here we go."

The first step in the program was to pump up the clients' often faltering egos. One by one, the eight women stood before the class opposite Didria. "Hello. My name is Brenda Fields," she began tentatively, practicing her interview skills. She seemed self-conscious, dressed up in a black skirt and pink rayon jacket, her hair pulled back in a tight French knot. Her nerves were out of order. What Brenda lacked in work experience she generally made up for with a strong sense of purpose. The program's director said that Brenda rated high in attitude and average in skills. "I'm here to interview for the food service job," she whispered.

"What did you forget?" Didria chimed in, softly. Flustered, Brenda held out her hand for the handshake. "Remember. Firm grip, in the groove. None of these finger shakes."

After each debut, Didria led the class in a "One, two, three—you go, girl!" Sometimes the women clapped their hands. Sometimes they stomped their feet. Sometimes she led them in both hand clapping and foot stomping. If the teacher forgot the routine, the women would remind her. It was their favorite part.

"Okay. Now look at me. What should I wear to this interview? A little skirt that's wider than it is long?" she asked, measuring her own waistline with her hands. The class laughed. Then each contributed to a list of fashion no-no's on the board: no dreadlocks, no big old ethnic jewelry, no new piercing, no visible tattoos, no loud perfume, no skirts slit up to your bottom, no bare legs, sensible makeup. The message was to blend in with corporate expectations, to wedge yourself inconspicuously on the ladder's bottom-most rung and only then to hope that someone will eventually notice you.

"Don't go in there looking like Bozo the Clown," said Didria. "I've seen some doozies." One woman nearly lost a corporate account for them because she showed up for an interview with ten different sex positions painted on her fingernails.

The motto at America Works was: the first job is the first step off the welfare stoop. The staff encouraged women not to dismiss a job automatically because the pay was too low, or the commute too long. America Works made its profit first by providing companies with welfare workers for four months, risk-free. It supported the worker with transportation costs and with counseling. In exchange for this service, it pocketed a portion of the worker's probationary pay, leaving a little more than minimum wage for the worker to take home. (The welfare recipient survived because social services continued to pay reduced benefits, Medicaid, and child care during this period.) The big payoff for America Works arrived only if and when its client was hired permanently. Then the state paid the agency a finder's fee from the welfare till—about $5,200 "per unit," in America Works speak.

The message worked on Brenda. She felt buoyed by the classes. The staff was pleasant. With Ty in home care, she used her time away from America Works to spread her résumé all over town. She traveled to more than a dozen hotels, including the Marriott in Brooklyn and the Novotel and New York Palace in Manhattan. She applied to work at a laundromat chain, at a Ranch One sandwich shop, at the Veterans' Hospital, at Chase Manhattan Bank, at Doral Inn. No one called. She attended a job fair for welfare recipients where most of the openings were for seasonal work. "One man with gold chains around his neck was hiring workers for a cruise ship," she said. "He was flirting with the young girls in front of me. When I got up there, he said there are no more brochures left for me. I told him, 'I guess nobody needs to eat on those ships of yours.' "

Like the majority of America Works clients, Brenda reached the end of her four weeks without a job and without prospects. That's when one of the sales reps tried to talk up a waitress position at the Red Lobster restaurant. "It was $2.95 an hour plus tips," Brenda recalled. "I told her I am too old to be hustling for my pay. I need something steady. That job was not going to help me get off welfare. And she told me, 'Well, Brenda, a job is a job.' "

In the summer of 1998, Brenda entered the most difficult part of her welfare journey. Her family's benefit check was cut to ninety dollars every two weeks. The city cut off child care help because her four-week job program was over. This was what the city called a "compassionate" nudge off the cycle of dependency. To Brenda it felt more like a slow, slithering noose. She pulled Ty out of Jesus Loves the Little Children. She could not afford the fifty-dollar-a-week fee out of her own pocket. On top of everything, her friend's seventeen-year-old son ran up a four-hundred-dollar bill on her phone calling porn numbers. She needed her phone in case an employer wanted to call. The culprit was J. J., the teenager who had shared floor space with Brenda and her children at the Emergency Assistance Unit at the beginning of this welfare odyssey. His parents, Alice and Lonnie, were struggling too.

One day stoic Brenda found herself dissolving in the director's office of America Works. "I had no money. I couldn't even buy a subway token to look for work," she said. "I had done something I never did before. I cashed in food stamps for money." Welfare recipients sometimes exchange a ten-dollar food stamp for seven dollars in cash at the right bodega or grocery store. The worst offenders buy drugs with it. "I paid the phone bill," Brenda said. "After that, I had nothing left." The director gave her three dollars in train fare so that she could make an interview for a cashier position set up for her at Paul Stuart, an exclusive clothing store on Madison Avenue.

Brenda called up a friend she had met in the Jackson Avenue shelter to see whether she could watch Ty. Deborah Johnson (not her real name) was a welfare-weary woman in her late fifties with ten children and a thyroid the size of a mango. She had subsisted on some form of public assistance nearly her entire life. Her current condition allowed her to collect Supplemental Security Income for the disabled. Deborah's third oldest had a crush on Brenda. Vincent moved gently, his jeans hanging loose off a lithe body worn down from years of fighting off asthma and cocaine addiction. On parole from his last drug conviction, he lived for a while at a Salvation Army rehab house. Later he worked at a pickle factory during the day, selling hats, gloves, socks, and jewelry on the weekends for pocket change. Vince came around so often to Brenda's apartment that Ty started calling him "my daddy." Ty looked forward to playing ball with the gentle vendor, or riding the trains selling socks. "Two for five, two for five," went the sales pitch. It was something Brenda discouraged but didn't stop. She was grateful for the break. Vince told Brenda his family never had money growing up. There were times when he and his siblings would go begging for quarters, hustling for bread and milk. His least favorite memories were the mayonnaise sandwiches and the lights. The lights were always on in the family's various apartments, to ward off the rats. Sometime in his teens Vincent turned to the fast-cash world of selling drugs. The money was good. The high was better.

Brenda began her ritual dress-up for the interview, wondering whether all this effort squeezing herself into thrift-shop clothes would ever result in a job. She was tired of painting the runs in her one pair of stockings with fingernail polish. She was tired of holding in her stomach. She was tired of wearing the same pair of shoes. ("I'm waiting for this pair of blue shoes at the thrift store to be marked down from $6.99 to $3.99.") Mostly she was tired of being disappointed. "I know I'm gonna get a job sooner or later, but this is stressin' me out." She tied a tasteful scarf around her neck, applied red lipstick, and bounded off, catching admiring glances as she walked toward Deborah's house with Ty, carrying his big lunch box, in tow.

The sight of Paul Stuart transported Brenda into a world where incomes were disposable, where subway tokens were a foreign currency. Recessed from the Madison Avenue sidewalk, the midtown store offered an intimidating display of fine couture. Shirts sold for as much as $1,900 apiece, shoes for $500. Salesmen with bow ties and palace guard postures were posted every other square yard. About half a dozen employees interviewed her that day, describing the job as if it were hers—six months to train on the register, $7.50 an hour, with union and medical benefits if she passed probation. She told the interviewers that she lived "by Yankee Stadium," that she had experience in retail at Alexander's, that she always had a good rapport with the customers. "I practically begged them for that job right there," Brenda said, prattling with expectations. The personnel director made an appointment for her to take some kind of a screening test. He gave her a slip of paper with an address. She traveled home imagining herself taking this route every day from then on, spending her working life handling the lustrous fabrics, making change for the highborn and hard to please.

The sight of Ty brought Brenda whirling back to the Bronx. He was playing by himself in the hallway outside Deborah's apartment when she arrived, dressed only in his underwear. He seemed deflated, not his rambunctious self. "I'm hungry" were his first words. "Didn't you eat the lunch I gave you?" Brenda asked him. "I had to share," he said, looking downcast. Brenda promised Deborah she would pay her when she got some money together. Then she asked, reluctantly, whether she could leave Ty again soon. Brenda couldn't imagine that Deborah would let any harm come to Ty. She was her friend. They both knew about each other's financial straits. They shared memories in the Jackson Avenue Family Shelter.

A few days later Brenda stood in front of the unmarked building on West Thirty-first across from Madison Square Garden in Manhattan, checking and rechecking the address. This was the right place, Sterling Testing. She was supposed to take her test here for the Paul Stuart job. Once on the elevator, Brenda turned her thoughts from Ty to the upcoming test. She figured it was a math exam, to see whether she could make change and do quick calculations. "Math I can do. But if there's any algebra on it, you can forget it," she thought to herself. The waiting room resembled a doctor's office. The dozen clients were hunched over clipboards and forms. A clerk behind a clear partition handed Brenda a written test and waved her back to her seat. She glanced through the booklet and saw true-false questions. None of them asked her how to add tax to a one-thousand-dollar bill. Instead, it was called an "integrity test." These questions were supposed to do nothing less than measure her moral judgment. She started filling in the answers, becoming increasingly uneasy. What was this for? Did Paul Stuart's executives assume she was dishonest just because she was poor? She had no idea.

Some of the questions were easy to figure out: "It's okay to drink alcohol on the job as long as it doesn't impair your ability to function." Or, "Anyone who has the opportunity to steal and doesn't is stupid." What idiot would get tripped up on those? Others posed philosophical dilemmas: "My personal freedom comes before any law." Were they screening for libertarians? Then there were the general statements that pitched Brenda into an ideological minefield: "Too many people are sent to prison," or, "No one can get rich without being dishonest in some way." Was it a moral failure to believe that prisons were overcrowded, or that the rich were sin-free? There was one question that seemed like an unavoidable trap: "What was the value of the last thing you stole? $1 to $10; $10 to $50, etc." There was no option for $0. She left it blank.

Next, she was called into a separate room for an interview. A pleasant enough man with a bushy mustache didn't reveal her score. Instead, he asked her whether she ever did drugs. Brenda said she tried marijuana a long time ago but didn't like it. He wrote "Marijuana" in big letters on the back of his sheet. He left it there, even after he discovered that Brenda was not supposed to take a drug test. Then he asked whether she'd ever been before a judge. A custody case, she answered, once when Ty's father tried to claim their son. "I mean, have you ever been accused of anything?" he asked. Brenda told him about an assault charge once lodged against her by a "trifling" neighbor. It was years ago. The charges were eventually thrown out for lack of evidence. He nodded, and made some notes.

Brenda left Sterling Testing feeling as if she had just backed into barbed wire. "I really messed that up," she said. "He threw me with those questions." She was angry at herself for telling the tester about that court case. It was sealed. She was acquitted. But it probably made her seem like a high risk. She couldn't get that big "Marijuana" scrawl out of her mind. One look at that, and the people at Paul Stuart would never let her cross its plush threshold. And that test. She felt like a criminal just by taking it.

Meanwhile, Brenda picked up Ty for the last time at Deborah's. She had a bad feeling as soon as she entered the apartment. She handed Deborah five dollars. Deborah turned the bill over in her hand and asked her if she could spare any more. Vince motioned Brenda to come talk to him in a separate room. He started telling Brenda that her lack of money was becoming a problem in the house. Brenda could hear Ty crying. She opened the door and couldn't see him. "Where's Ty?" she asked. The girls laughed and said, "Oh, he's outside." Brenda opened up the door to see Ty sniffling in the hallway. She let him in, and the young teen told him to "stop crying" and whacked him on the back of his head. Brenda lunged for the youngster and pushed her backward. "That girl has no business putting her hands on my son. He's a baby," Brenda bellowed. Cursing and screaming spilled out into the hall. Brenda called the police. She and Vince gathered up Ty's things and bolted out of the building.

Tensions escalated as the evening progressed. Brenda called Deborah to ask her why their friendship ended like this. Deborah's daughters called to berate Brenda for manhandling their thirteen-year-old sister. The phone rang all night. Ty tossed and turned with nightmares. He sat up and screamed in his sleep for his mommy. With his eyes wide open, he kept calling and calling. "Don't! Stop it!" It took her hours to calm him.

In the ensuing days, the accusations crescendoed into hysteria. Vincent accused another sister of threatening him with a handgun. An anonymous caller left a message on the America Works phone, saying they "had a thief in their midst. Brenda Fields. She watched my

daughter and took my VCR and my TV. You'd better get rid of her."
The staff ignored it. Brenda requested an unlisted phone number. She took out an order of protection against the family, charging aggravated harassment.

A week later Brenda was frying chicken in her apartment when a surprise visitor showed up at her door. It was a caseworker from the Administration for Children's Services, the city's child welfare agency. Ty was playing with his train set in his room. Loreal was listening to hip hop music in her room.

Brenda asked the agent whether he wanted to inspect her refrigerator. She knew that was routine for child welfare agents, to see whether the mother was providing food for her children. Brenda had received her food stamps that morning. She spent all day stretching $207 into groceries for a month, going from store to store to get the best buys with her coupons. A fresh watermelon was sitting on the table. Her freezer was full. "The agent said it was okay, but I opened it up for him to see anyway." She showed the caseworker the chicken she bought for thirty-nine cents a pound and the pork shoulders she bought for fifty-nine cents at Western Beef. Under the sink, she showed him a ten-pound bag of rice for $4.99 and an industrial-sized jug of Mazola Oil from Pioneer Grocery. Cans of corned beef hash, jars of applesauce, and a twelve-pack of generic soda for $2.99 were still waiting on the counter to be put away.

Brenda kept cooking while the caseworker told her the reason for his unannounced appearance. Someone had left an anonymous complaint with ACS about Brenda, Vincent, and Loreal. The caller said all three adults did drugs every night in front of Ty, and that Brenda sexually abused her son. The caller also said Vincent beat the four-year-old. Brenda was unusually calm as she listened to the string of horrors. "I knew right away why he was here," Brenda said. "Ghetto people are famous for doing stuff like this—calling BCW (Bureau of Child Welfare, the old name for ACS) to harass one another." She figured this was another volley from the Johnsons.

The investigator asked Ty to show him his toys. Ty took the man into his room and showed him his favorite train. "What's your favorite TV show?" he asked Ty. Power Rangers. "Favorite food?" Raviolis. "Did anyone ever touch your private parts?" No. "Did anyone beat you?" No. "Do you want to stay with your mommy?" By this time Ty was sitting on his mother's lap. He hammed it up with a big smile. "I loove my mommy," he said, giving Brenda a wet kiss on the cheek. The caseworker left, saying he was required to make one more visit. After that, he would close the case. "He told me the moment he stepped in the door he could see what's what."

After a failed attempt later to settle their scores in mediation court, the dispute between Brenda's family and Deborah's finally subsided, leaving a friendship in tatters and a four-year-old boy haunted by night terrors. "I should have known better than to get involved with them. I don't like to say I was desperate, but I was desperate," Brenda said. "I had no one to depend on for baby-sitting except Deborah. I thought she was my friend." Another miscalculation.

In the midst of the trauma the director at America Works told Brenda she didn't get the Paul Stuart job. The personnel manager insisted it had nothing to do with the integrity test. They just found someone more qualified.

A few days later Brenda found herself dashing down Wall Street, running slightly late for another interview. This July day marked nearly ten months since she had moved out of the shelter. She'd been looking for work ever since. She began thinking her time would have been better spent learning some new skills, or getting herself an associate's degree in medical assistance.

It was a glorious summer day as Brenda emerged from the Rector Street subway station. She glanced up at the odd vision of Trinity

Church to get her bearings. It stood in dark red Gothic serenity at the tip of the world's most frantic financial street. Wall Street was a winding pathway, named for the structure built by African slaves to ward off attacks by Indians and Britons against the Dutch businessman Peter Stuyvesant's settlement. In the months before the millennium, Wall Street was considered the most prosperous address in the world. Messengers, flower deliverers, and cooks shared its early morning pavement with tourists, investment bankers, and stockbrokers.

Brenda made her way past the colonial Federal Building where George Washington was inaugurated, past the nineteenth-century Merchants' Exchange, to the massive granite columns of J. P. Morgan headquarters. Loreal was home with the flu, watching Ty. Brenda was reporting to another job interview, this time in a basement. She had a little more than three years left on her welfare clock. With each rejection, the ticking had grown louder.

Two weeks later some good news finally arrived. Brenda got the job. She was hired to work a forty-hour week feeding the employees at J. P. Morgan, working the register, stocking the gourmet salad bars, and operating the cappuccino machines. Her employer was Aramark, a $7 billion global corporation recently subcontracted to operate the Morgan cafeteria. America Works had a longtime relationship with the catering company, which had also just sealed the deal to provide meals for the 2000 Olympics in Sydney, Australia. For four months Brenda would receive minimum wage while welfare continued to pay her transportation, child care, rent, and health benefits. America Works called it "supported work." Then, if Aramark liked her, the job would be hers. She would sever her ties to welfare.

Work was going to save her.

9

Ruddy Evangelist

Jason Turner flew in from the frigid beer town of Milwaukee, Wisconsin, for his chilly New York City debut on a January afternoon in 1998. The burly, forty-four-year-old new city welfare commissioner was leaving behind the nation's most radical welfare reforms to take on the nation's most daunting social service system, a thousand miles east. He brought with him the rumpled composure of a midwestern stoic, and a family for moral support. Angela, his wife, was seated nearby in the Blue Room at City Hall, cradling the couple's newborn baby, Sarah, their fourth child. Turner looked out over the gathering of reporters. His half-smile would become an inscrutable trademark expression. He needed all the camouflage he could muster against the coming artillery attacks.

Turner's presence as commissioner marked a directional shift in the winds for welfare's future in the city. A policy that had been strictly WEP-centric, with City Hall at the helm, would now change tack. The agency's programs were to become more varied, more creative. Welfare supervisors in the trenches would be called upon to lead these revisions. The agency's message would become ideologically purer. The grandson of a banker and the son of an advertising executive, Turner was not a professional government bureaucrat with an eye on the city's career ladder. Nor was he a commissioner-come-lately to the welfare reform scene. In his perfect world, Turner would imagine away welfare in its entirety rather than alter it. His core philosophy was in sync with that of the nation's most recent welfare reformers. But beyond that, Turner was a true believer, weaned as a young man on a firm conviction in the social contract, inspired into adulthood to find ways to render the public dole obsolete. Overhauling the welfare state was very nearly his religion.

"For somebody in the welfare community, New York City is it," Turner said, stirring some relief among those gathered. Here was the proper show of deference to New Yorkers, who never failed to be impressed with the size and scope of their own problems. He was correct in a demographic sense too. White and rural recipients had already fled the rolls in disproportionate numbers nationwide, leaving a welfare population that was browner, blacker, and even more concentrated in the largest cities than ever before. So New York City was, in fact, it. "This is the place where it's possible to make large amounts of difference in many people's lives," he said.

A natural skepticism preceded Turner's abrupt arrival. New Yorkers often imported such outsiders but rarely trusted them for the long haul to run their government agencies. The Midwest to most New Yorkers was a hayseed's jumping-off point, somewhere between Forty-second Street and Hollywood. How could anyone west of the Delaware Water Gap fathom the complexities of the city's racial politics, its massive size, its delicious idiosyncrasies? A leader who learned his craft in the vague vicinity of a cow pasture was quickly dismissed as too innocent to survive.

Besides, Turner was an unreconstructed conservative, far to the right of the mayor who hired him; he was a Reagan Republican in the heart of America's famously left-leaning city. Dodging the missiles lobbed by New York's liberal advocates for welfare, the homeless, hunger, housing, and AIDS victims would occupy a great deal of Turner's energy.

Turner was prepared for the challenge, though he preferred, of course, the company of like-minded conservatives. In Wisconsin he had honed his ideas with policy planners at the Hudson Institute, a group that counted former Vice President Dan Quayle and General Alexander Haig, Richard Nixon's former chief of staff, as board members. The common goal of the Indianapolis think tank was to promote the power of work to the top of the welfare priority list. Its plans included privatizing job searches and discouraging single parenting. But most emphatically, the social welfare thinkers hoped to engage every public aid beneficiary in some form of activity in exchange for their check. The group's policies were already well entrenched in Wisconsin's welfare system by the late 1990s. By the time Mayor Giuliani asked Turner to consider taking on New York's welfare population, already 70 percent of the recipients in America's Dairy State had left the rolls, a drop so steep that it surprised even Turner.

But Wisconsin was a mere warm-up for the hurdles hobbling welfare reformers in New York. In the midwestern state, Turner faced the challenge of putting thirty-one thousand single mothers to work in a state with a 3 percent unemployment rate. New York had more than ten times that number of women on welfare, with an unemployment rate nearly three times higher. So heading up the Human Resources Administration in New York City was not just a $137,000 per year job to Turner. This would be the ultimate practical exam for his homespun theories. His very survival under Giuliani's regime would be the most telling test of his colleagues' ideas, of the congressional edicts embraced two years earlier. If this welfare expert could eliminate the handout in New York, he could eliminate it anywhere. "This was the golden opportunity of a lifetime," Turner would say in a later conversation. "It's something that strikes only once, like lightning."

If Turner was a dedicated policy designer, he was also a ruddy-faced evangelist, summoned to convert America's secular capital to the fundamental generosity of work. He never gave up hope that a civic majority would eventually embrace the wisdom of his views. "Work is society's organized way of giving gifts to others," Turner said in a June 1999 interview. "It fulfills a basic human need. It links the individual to mankind. If you set up a system where you accept income without work, then the link is broken. So the redemption from work comes in reconnecting that link. You restore yourself economically by working, and you restore yourself spiritually."

Occasionally Turner's zeal to pass on his beliefs caused him to trip into the city's ubiquitous hostile waters. This was an overwhelmingly Democratic town, by voter registration, if not inside City Hall. Its governing officials tended to bristle when lectured on philosophy. They wanted plans, details, quantifiable results. One of his first appearances before the city council's General Welfare Committee foretold an explosive relationship with the lawmakers empowered to oversee his vision, and with the welfare recipients obliged to live it.

The new commissioner's debut in March 1998 happened inside the ornate hearing chambers at the top of City Hall's colonial spiral stairs. The room was packed. Word had spread throughout the advocacy community that the new man from Wisconsin was prepared to add more procedures to their already rule-bound lives. Close to two dozen welfare recipients, many dressed in paint-spattered boots and hooded sweatshirts, filed into the chairs behind Turner's table. Dozens more were blocked outside City Hall's gates, barred from entry by the mayor's new orders to limit public access to the seat of government, for fear of foreign terrorism. This ad hoc group was arguing for safer working conditions, real wages, and a union for WEP workers.

Turner began to read his opening statement, keeping eyes on the text, unfazed by the agitation behind him. "Mayor Giuliani wants welfare to cease being a matter of cash assistance without obligation," he read, popping cough lozenges into his mouth. "Work in return for benefits is always better than doing nothing for money." Turner spoke over the jeering crowd, outlining his leitmotif: turn welfare centers into job centers, and require work for everyone, "with very, very few exceptions." WEP would no longer be the only game in town. It would be a sidebar to a broader plan to engage 100 percent of the city's beneficiaries in work. WEP was pretend work, he figured, "simulated work," as he called it, good for building character at least, if no other activity could be found.

Turner's soon-to-be nemesis, the General Welfare Committee chairperson, Stephen DiBrienza, openly mocked the new commissioner's presentation as "the view from *your* world," a view long on philosophy but short on details. The brash Democrat from Brooklyn scolded Turner for failing to provide copies for the members and for being unprepared with answers to their specific questions. With the roll of his eyes and the crack of his gavel, this sarcastic politician with shellacked hair sent the new kid back home to get his homework.

It was a rancorous beginning, but one that did not surprise Turner. He had expected an onslaught of ill will. He was well aware that he was the enemy in this town of "intellectual elites," as he sometimes called them. Frankly, he seemed to flourish in his bunker. He was convinced from his years in government work that the remedy for poverty was not to throw the public's money at the problem, but to craft new rules of personal behavior, and to enforce them. DiBrienza represented the opposite view. He had come to believe that welfare reform, at least reform Giuliani-style, was little more than a prod to keep the underclass under toe.

The two men's relationship would become even more embittered over time. Both represented extreme partisan differences, in substance and style. Neither Turner nor DiBrienza could contain his open hostility for the other. DiBrienza would rail on the record, calling the commissioner "arrogant, disrespectful, evasive, disingenuous." Turner would respond by storming out of the hearings, or failing to show. At one hearing Turner bolted from his chair, complaining that DiBrienza's famously loud voice was too loud. As the audience whooped and clapped, DiBrienza bellowed to the departing commissioner: "Welcome to New Yahk! This is how I tawk."

Compared to upcoming storms, this March hearing was a fairly innocuous squall. When talk turned to adjourning early, some of the Wep workers stood up in their seats, chanting, "Union now! Union now!" A few held handmade signs. DiBrienza banged his hammer for order. Security guards escorted the welfare recipients out the door. One protester wheeled around for a parting shot: "Cheesehead!" he hollered, a reference to those silly wedges of foam cheddar some Wisconsin football fans wear on their heads. "Cheesehead! Cheesehead!" others echoed. Turner's neck flamed over his white shirt. Then he turned around to get a look at his tormentors, and smiled his half-smile.

This was not going to be easy.

Some three months later, in a June public television interview, Turner, still the newcomer, strolled unwittingly into one of New York's many ethnic bogs. During a wide-ranging discussion that veered from welfare to work, the host of *Thirteen On-the-Line* asked the new commissioner what he thought of union leaders who said the city's Work Experience Program was little better than slavery. Turner looked puzzled, and not a little perturbed.

"You and I are sitting here, and we both put in a long day today. Are we slaves?" Turner said to the host, Brian Lehrer. "I would say everybody has an obligation to work. In fact, it's work that sets you free."

Work sets you free. It was a variation on the slogan made infamous by the Nazis, *Arbeit Macht Frei*. The same inscription hung over the front gate of Auschwitz, Hitler's horrific attempt to fool the Jews who passed underneath into this factory of mass genocide. Death was shrouded under the mirage of forced labor. It was the most injurious image Turner could have invoked. It insulted the mayor's prized political constituency, New York Jews. Maybe worst of all, it suggested that work might not be redemptive but manipulative. After viewers called in to complain, the commissioner released a written apology. His intention was not, he said, to compare the experience of workfare for blacks and Latinos to the horrors of the Holocaust for the Jews. "It was my intention to state the view that work can give an individual a greater measure of personal freedom, independence, and self-sufficiency in contrast to the dependency of welfare."

The mayor stood behind his commissioner. Turner's opponents would never let him forget.

I f work was Turner's central message, it was also his own personal code. In Wisconsin the dedicated bureaucrat was known to bunk down on his office floor in Madison rather than make the eighty-mile commute home to his wife and children in Milwaukee. Later, in New York, when Turner appeared at public events with crinkled suits and flyaway hair, rumors soon spread that once again the working dervish was camping out in his office. But whether he spent the night in the newly renovated HRA executive suite or his Queens condominium, the point was undisputed. Turner practiced what he preached. He worked nonstop, pouring sleepless months into the task of changing the culture of welfare.

Most public servants who make poverty their life's work are inspired somehow by personal experience with the poor, a Peace Corps encounter, a parent in social work. Turner's inspiration was more cerebral than practical, honed, in part, from a life lived far away from poverty's margins. Born John Anthony (a name he changed to Jason after college graduation), Turner was the son of a homemaker and an affluent advertising executive whose firm marketed Crest, the toothpaste, and Zest, the detergent. He grew up in Darien, Connecticut, a town of yacht and curling clubs, where the average home cost nearly $900,000. From his home in Darien, welfare was an abstraction, even an aberration. It rattled Jason's notion of how people should behave.

The first seeds for Jason's view of the world were planted in the United Kingdom. There, as an expatriate child, grammar school provided him with a youthful taste for propriety and obligation. England's exacting rules of behavior, its notion of place and decorum, provided a comfortable space for a child who craved order. "The social system there was structured, very hierarchical," Turner remembered. Children knew how to act toward one another and toward adults. The social code was quite clear. By contrast, American-style fifth grade struck the returning ten-year-old as "chaotic." Too much random behavior, too little deference to authority.

At age ten, Jason took a cross-country tour with his grandfather, a dignified man who helped mold the boy's evolving work ethic. John Tufel impressed his grandson not so much with his travel skills but with his thrifty budgeting and tales from the leaner days. When the Great Depression hit, Tufel lost his job as a Wall Street stockbroker, joining the armies of the unemployed. "Like many people, he would have done anything possible to avoid relief," Turner said. "And he did." Stoic effort meant everything, not timing, not educational background, not the greater economic conditions. Grandma did her best to keep their Jamaica, Queens, home running as if little were awry. She insisted that her husband put on a suit every day, even if he had nowhere special to go. Looking for work became his grandfather's job. At the end of the day Mrs. Tufel prepared a hot dinner.

"Grandma would never let him sit at home and feel sorry for himself," Turner said.

Turner's grandfather made ends meet by selling Fuller brushes door to door. The job was commission work, not salaried. Big sellers earned prizes from the company, like sets of dishes and napkin holders. It was enough to sustain the Tufels. When the bad times ended, Grandpa resumed his place in the financial world, becoming a bank president in Waterbury, Connecticut. "He was a proud man," Turner said. "A man of firm principles."

The epiphany for the future welfare whiz came when he was twelve years old. At an age when other boys were reading comics or planning pickup baseball games, Jason was reading news magazines and worrying about a society corrupted by unrestrained generosity. A 1965 story on welfare abuse in *U.S. News & World Report* set Turner on his future course. Entitled "How It Pays to Be Poor in America," the article included a list of the freebies the poor could collect without working for them: cash benefits, food stamps, public housing, medical care, schools. Jason's understanding of social order was turned upside down. "I thought then, what if everybody did this? Collect money without working?"

As a teenager, Turner worried about this specter of "enforced idleness," as he would later call it. He spent much of his spare time during his high school years designing fantasy factories for the poor to make simple things, like Christmas ornaments. It was his own WPA-style project for the sixties, for those he thought needed to do something useful in exchange for their government gifts. "An opportunity to work," he believed, would solve the problem of poverty.

Turner veered off the anti-welfare career path as a student at Columbia University. He enrolled as an undergraduate history major three years after the Ivy League school's turbulent student protests of 1968. Conservatism was a lonely path for a college student at the time. Turner fell in with the class sentiment and voted for George McGovern for president in 1972. But the nascent conservative was abruptly reawakened during a college break in Houston, where he was twice robbed at gunpoint while driving a cab.

By 1980 Turner found himself stumping for Ronald Reagan's presidential campaign. He managed an appointment as a federal housing official in the Reagan regime. He quit in 1985 to try his hand at turning a personal profit as a landlord in a D.C. ghetto. Turner set up Czar Realty and sank his savings into twenty-five apartments that were soon engulfed in the local crack trade. After three years, with his rental empire belly up, Turner returned to government service in 1989 as a senior welfare official in the George Bush administration.

Turner was unable to find a niche for his bold welfare changes in the Bush White House. Still, he stuck out his term, and left only when Democrat Bill Clinton was voted into office. That's when Wisconsin Governor Tommy Thompson captured Turner's attention. Like several other Republican governors, Thompson was experimenting with reinventing welfare from the ground up. Wisconsin had dismantled Aid to Dependent Families with Children three years before the federal government recommended the same on a national level. Individual benefits were reduced. Job search requirements were increased. Wisconsinites began leaving the rolls in rapid numbers.

The northern state had become a ripe canvas for conservative ideas. Turner was so enthralled by the possibilities there that he agreed to take a lower-level bureaucratic job in the Thompson administration just to be part of it. With help from consultants at Indiana's conservative Hudson Institute, Turner then helped draft a tough welfare plan called Wisconsin Works, or W-2. He was now able to tinker with the behavior of an economic class of people. "It's plumbing the soul," he told the *New York Times'* Jason DeParle, "figuring out why people do the things they do."

Wisconsin Works was at once punitive and generous. If welfare recipients behaved correctly and followed the tough work rules, rewards would follow. Every aid seeker worked from the moment he or she applied for assistance, either in private industry or in community service jobs. No exceptions. For those who violated the work rules, benefits were docked, then terminated if they continued to slack. Again, no exceptions. On the other hand, lawmakers raised the standard benefit to $8,100 a year per family. They invested heavily in health care, child care, and wage subsidies for those who left the rolls. They added vocational education programs and day care subsidies linked to income.

By the time Turner was mulling over a move to the New York job, the Wisconsin poor had left public assistance en masse, but little was known about their fates. Like New York, Wisconsin did not systematically track the futures of those who forfeited their welfare checks. One informal survey in 1997 found that the year Wisconsin dropped 43 percent of its welfare recipients, its homeless shelters in Milwaukee experienced a 30 percent hike in people seeking a bed. Food pantries received 90 percent more phone calls.

A later University of Wisconsin survey of families leaving welfare found that 68 percent were working after they lost their benefits. But work didn't necessarily translate into a decent living. In fact, most of them were earning about four hundred dollars less than they would have if they'd stayed on the dole. Only one-quarter of them had lifted their families just over the poverty line. The state's Department of Workforce Development calculated that 68 percent of those who left the rolls reported they were "barely making it." More Wisconsinites than ever—30 percent more in 1997 than 1989—were living in extreme poverty. Another study found that infant mortality rates rose steeply, just as food stamp participants declined. Turner did not like these trends. Still, he believed the benefits people gained by being free from welfare would eventually overshadow this temporary pain.

• • •

When Turner rode east to head up HRA, the public was still in the thrall of the amazing arithmetic of welfare reform. Forty-four percent of the nation's welfare recipients were gone since the peak of 1994. Fifty-six percent of Texas recipients had left the rolls, 63 percent in Colorado, and 80 percent in Idaho. For sheer percentages, no state topped Wisconsin for shaving its public assistance roll down to a sliver of its former self. Eighty-two percent of Wisconsin's poor had left public assistance by 1998. (By the turn of the century that number would increase to 93 percent, down to seven thousand recipients from one hundred thousand five years earlier.) Turner was the chief architect of this mathematical phenomenon. Mayor Giuliani hoped he would roll out similar numbers in New York. After he worked his magic in the city, political insiders believed, Turner might leave for a White House post in the next Republican administration, particularly if his Wisconsin mentor, Governor Thompson, was chosen to head the federal Health and Human Services Department.

Many HRA employees were relieved when Turner arrived. Bullying from City Hall aides and the rush to WEP had dominated the last four years of their workaday lives. Staffers were looking forward to a leader who prodded less and theorized more. "Until Turner came, all we heard was 'Close their cases no matter what,'" said Marilyn Moch, who would retire as HRA's director of training and procedures in 1998. "At least Turner might return some more programs to the mix."

The first thing Turner did was try to shore up the eighteen thousand troops in the field—the caseworkers, income support center supervisors, the frontline staff. He arranged for en masse briefings (or as one HRA employee called them, "ideological harangues"). Over a six-day period, streams of staff from the various boroughs came in shifts to the quaint Town Hall auditorium on West Forty-third Street to absorb the new vision. The sessions were called, rather clumsily, the

"Toward Welfare Reform Culture Change Training Program." Emphasizing self-reliance for welfare applicants, the program aimed to change "unproductive behavior" and save taxpayers money. The message was: every adult is capable of work.

"Frankly, most of the staff appreciated those sessions," said Moch. "No other commissioner had paid attention to them. He understood their level of work was important." HRA has never been known over the years to be client-friendly. The new administration tapped into a culture of welfare staffers who had believed for years that enforcing a work ethic was the only way to go.

As Turner threw himself into changing the welfare culture, he left a large part of himself back in Milwaukee. Angela stayed behind with their four children in the couple's Shorewood, Wisconsin, home to run the for-profit consulting firm she and Jason had founded, the Center for Self-Sufficiency. Jason flew home to see them nearly every weekend, adding jet lag to workload. But at the same time Turner brought with him a team of familiar faces from the group of Wisconsin Works architects.

Andrew Bush, his boyish-looking program analyst, would become the new New York deputy commissioner for policy. Bush had worked side by side with Turner drafting the shape and scope of Wisconsin's reforms. He went on to direct the Welfare Policy Center at the Hudson Institute, the Indianapolis think tank that was funded in large part by the Bradley Foundation, a wealthy Milwaukee-based contributor to a wide range of conservative social welfare causes.

Marriage was to Bush what work was to Turner—the key ingredient in combating poverty. Bush argued that the absence of fathers, perhaps even more so than the absence of work, kept poor families whirling in place. "There is no greater single threat to the long-term well-being of children, our communities, and our nation," Bush wrote in 1997, "than the increasing number of children being raised without a committed, responsible, and loving father." Welfare, he believed, encouraged dads to flee. Bush believed that a new welfare system could help regulate people's sexual mores by funneling prime aid to heterosexual couples who were legally married.

New Yorkers, like most Americans, never really embraced the parts of the federal welfare act that legislated the size of poor families and their sexual mores. Only a handful of states, New Jersey among them, put caps on the number of children welfare mothers could support with public assistance. "The consensus is on laws that focus on behavior outside the home, not inside," said New York University political scientist Lawrence Mead, who also became a consultant to the Turner administration. The author of the 1992 *The New Politics of Poverty* had helped the Hudson Institute evaluate Wisconsin Works. "No one has really found a way to legislate traditional families," Mead said. "There is no government solution to single motherhood. But there is a consensus that the poor need to follow dominant values, to work and send their children to school."

If Turner's mantra was work, and Bush's marriage, then Mead's was regulating the lives of the poor. "It is the return of social control," said Mead in a soft, gravelly voice, referring to his overarching goal of welfare reform. Mead spoke from his NYU office, shades drawn, door closed during a May 1999 interview. Some of his students, he said, had just complained to the department about his class, accusing him of spreading racist ideas. Mead was about to face a tribunal of his accusers on the department's orders, to hear their arguments. This was clearly something he was dreading, a showdown on race, a turbulent issue he considered a distraction from his main point.

"Students jump to the conclusion that a statement is racist without confronting the facts," Mead said. In his view, the issue at hand was a contest between the rational and the emotional. "I say it's a rational response for a white person to cross over to the other side of the street when he sees a group of black teens walking toward him. Black teens are in jail at a higher rate than any other teen group for violent crime," he said. "It's rational. It should be up for debate. But no, to students, it's only racist.".

Mead envisioned conservatives like himself in the bunker with Turner, while the "intellectual elite"—university researchers, media, lawmakers—launched salvos from above. The entrenched left-wing opposition, he said, was the main reason why administrators were loathe to release much data to the public. Such reports would be used against them, by the politically correct, in the heat of ideological warfare. Journalists and researchers had long complained that the Giuliani administration refused to release benchmark data to back up its claims. Hard facts on the fate of the half-million former recipients might have quelled the dissension. But this was war. And in war you don't always announce your casualties.

Within weeks of his arrival in New York, Turner and his team had already launched revolutionary changes. There was Mark Hoover, a thirty-two-year Wisconsin bureaucrat with a genius for implementing policies and a penchant for wearing cow socks in public. The other midwestern teammate was press secretary Debra Sproles, the only black among Turner's Wisconsin colleagues.

The first order of business was to convert the thirty-one income maintenance centers into job centers. "It's more than a name change," Turner told the city council. Job centers would be "places where all applicants for benefits look for work," he said. They would be laboratories for his ideas, the arenas where the gospel of work could be enforced, regulated, and spread. The key difference between the old and the new would be up-front job searches. Applicants would be required to hunt for employment for a month or more before their cases could be opened. This was the linchpin of his plan. Work would truly be real, not just a slogan.

Turner installed "Job Center" signs outside the usually anonymous welfare centers. He repaired and repainted the grungy offices so they looked more inviting, more professional. Inside, simple posters reminded applicants that public assistance was no longer a possible way of life. Framing a kindergarten-style clock were the words: "This Clock Is Ticking and So Is Your Time." On another poster the words "Welfare Is Time Limited" were superimposed over a tilted hourglass. To underscore the message, a receptionist would make periodic announcements reminding applicants of the new deadlines.

The old application clerks were given new official titles ("financial planners") and unofficial ones ("diversion experts"). Their job was no longer to sign up applicants but to try and find ways for people to avoid the burden of welfare. Did they have relatives who could help? Were they eligible for federal welfare, such as veterans' benefits or Social Security? Could they go to a food pantry to avoid getting on food stamps? The point was to counsel people off the city's rolls before they ever got on.

If diversion didn't work, the applicant was then sent on to the "employment planner," who would write up a schedule of eight-hour-a-day job searches. Families with children would have to search for thirty days. Single adults would search for forty-five to fifty days. If, at the end, they were still unemployed, the applicant could then receive public assistance, along with an immediate WEP job. Private job placement companies organized new groups of applicants into orientation sessions, giving them motivational videos and sending them off on job interviews. At the same time the client was required to face a series of other appointments, with the fraud detector, the finger printer, the eligibility interviewer. One missed appointment without an excuse could send the applicant back to square one.

The job centers represented a radical shift in the way welfare had always done business in the city. Turner was pleased with the first wave of reforms. Calling them "beautiful systems of engagement," the commissioner sketched out plans to have all thirty-one offices converted by the following spring. "The best time to reach people is not after they've been on welfare four, five, or six years, but when they're applying," Turner told *Newsday*. "You're still at the point when you can motivate them to consider alternatives."

Speed was part of Turner's strategy. The first one, Greenwood job center in Brooklyn, opened one month after he took office. The second opened in Jamaica, Queens, three weeks later. Two more opened in July. Nine others were converted in 1998, creating at least one for each borough.

"I thought it was going to take more time to roll out the job centers than it actually has," Turner told a group of lawmakers and social workers that November. "We didn't do lengthy planning. Instead, we acted first and worried about consequences later. And that seems to have worked for us."

Yet speed brought with it some inevitable casualties. In the early months the first Brooklyn job center turned 69 percent of the applicants away at the door. The Queens center sent 84 percent home without an application. The Concourse center in the Bronx sent home nine out of ten applicants over a four-month period.

Legal aid attorneys were hearing more and more complaints about people being denied emergency help they asked for. Teen parents with no money were being told they were too young to apply. Women were sent home without an application because their spouse wasn't with them.

But if the rush to reform caused some reckless consequences, Turner figured the bugs would be worked out. "There are all kinds of complexities in operating a large system," he said. All the grumbling would die down once the culture of work was given its chance to take hold. In mid-November 1998, Turner told a group of Albany policymakers that one role of the welfare system was to "create, if you will, a personal crisis in an individual's life." Time limits helped cause a sense of urgency. So did refusing to help tenants who owed their landlords overdue rent. Hardship would fuel compliance, which would lead to financial independence.

Turner's opponents hunkered down.

he Albany speech became an accidental turning point in Turner's tenure. Public approval began to falter. Turner's changes were coming too fast, and they were too severe. The only report his administration released in 1997 that tracked the fates of those who left the rolls was considered so unreliable that the city council derided it and the media virtually ignored it. The federal government became alarmed by the harshness of New York's policies. Medicaid officials began investigating complaints that HRA had denied emergency health insurance to the poor. The U.S. Department of Agriculture looked into complaints that the city was withholding emergency food stamps from hungry applicants. Both later found that the city had violated the law in denying assistance to eligible applicants.

The federal rebuke slowed Turner's momentum to a crawl. The very government that had enacted the strict work rules was now scolding New York City for taking them too far. At first Turner tried to ease restrictions on food stamp distribution, but the mayor cut him short. Giuliani dug in his heels, accusing some food stamp seekers of trading in their stamps for drugs. The city's legal aid attorneys began building a voluminous class action suit against the job centers. Lawyers in the field collected testimony from clients who had been turned away, hat in hand.

Lakisha Reynolds, a twenty-five-year-old mother of a three-year-old, testified that she tried to get emergency food stamps and cash at the Hamilton Job Center in Manhattan in December 1998. Lakisha had lost her job months before. Her rent and electric bills were overdue. She had one dollar in her pocket. "The only food I had left," she said, "was a piece of meat, two cans of vegetables, three tangerines, some hot cereal, a little rice, and some carrots." The first financial planner sent her home with orders to look for a job and a voucher for a church-run food pantry. The pantry closed before she could get there. The next day she tried, the pantry had run out of food by noon. The third time she went early to stock up on supplies but was late for a job center appointment as a result. The receptionist told her she had to start her application over from scratch.

April Smiley, a seventeen-year-old mother of one, testified that she asked for emergency help at the Jamaica Job Center in Queens just before Thanksgiving in 1998. She'd been bouncing from one home to another after her mother kicked her out because of the pregnancy, and her father in Florida stopped sending aid. She had no permanent home, no income. Her one-year-old, Anaysha, had no milk, no winter coat, and only one pair of shoes. "I only have one diaper left," April said, "and am worried about how I am going to be able to get more for my child." The caseworker told April she was too young to apply for any assistance and sent her home.

Thirty-nine-year-old Lue Garlick said she was homeless and four months into a high-risk pregnancy with twins when she sought help at the Hamilton Job Center in Manhattan in late November 1998. The receptionist told her this was "not a welfare center. It was a job center," Lue testified. The financial planner scheduled Lue for fifty days of job search duty before she could receive food stamps or Medicaid. Sick and dizzy with anemia, Lue said that she kept telling the workers she needed to eat. The caseworker referred her to a local food pantry, which was always closed by the time she finished her job center duties. In the meantime, Lue lost her rented room. She had no money. Her grandfather was too ill to house her. So she had to sleep in the park on 146th Street and Lenox Avenue. She would go days without eating, weeks without the medicine she needed. "I am desperate and have no place to stay from day to day," Lue said in her affidavit. "I am pregnant and feel sick all of the time. During the day I have to wander around.

The little food I get from my friends is not enough, and I often go hungry. . . . I do not know what to do anymore."

Four others recounted similar experiences for the same class action suit against the city. A pregnant nineteen-year-old who was trying to finish high school was denied health insurance or food. "A friend lends me one dollar each day for lunch," said Jenny Cuevas, "with which I buy a roll." A twenty-four-year-old woman who was nine months pregnant was told she could not get food stamps because she was not a citizen. A forty-year-old homeless man who was hit by a car spent his last $1.50 on the subway trying to get the documents required by a Harlem job center. After more than a month he still had no food stamps. "When I got to the food pantry on my own," said Elston Richards, "all they had was some old red bananas." All six applicants were told, erroneously, that emergency assistance no longer existed. All were sent to job searches hungry, or sent home empty-handed.

The city lawyers argued that these cases represented isolated, anecdotal, and tragic errors that should not be used to condemn an entire system. HRA was attempting a grand-scale overhaul of a monumentally flawed system. Mistakes and setbacks would inevitably occur. The city agency would do its best to make sure these errors would not happen again. But Federal District Court Judge William Pauley III warned of a deeper problem. The judge noted that the vast majority of people were turned away from all thirteen job centers without receiving help. "Work first," he wrote, had become "divert first" without mercy. The judge cited city statistics that of the first 5,300 people to go through the new job search program, only 256 were placed, an employment record of just under 5 percent.

Just before Christmas 1998, Judge Pauley ordered city workers to stop the new job search procedures. He gave the city time to mend its ways. He called a halt to any more conversions, stymieing Turner's plans to transform the city by the following spring. HRA spokeswoman Debra Sproles called the suit "another attempt by the dependency industry to keep the old, failed welfare system alive." Giuliani accused Judge Pauley of being mired in the "old philosophy of dependency that comes out of the 1960s and 1970s that helped destroy lives, not help people."

A month later the judge, a Republican appointed by former Senator Alfonse D'Amato, landed a second blow. Judge Pauley ruled in January 1999 that the city had overlooked the "urgent needs" of the poor in its zeal to reform welfare. Pauley said these new welfare policies were literally putting people's lives in danger. The Supreme Court, he noted, had already decided that denying welfare benefits to eligible applicants deprived them "of the very means by which to live." After a year of making some adjustments, the city welfare administration "continues to endanger numerous individuals in need of public assistance including children, expectant mothers, and the disabled," wrote Pauley.

Still, Turner found some hope in the court decision. The judge had praised the city's ongoing efforts to fix their problems. "This Court is impressed by the conscientiousness of Commissioner Turner and his deputies in their efforts to modify policies and procedures in the job centers," Pauley wrote. "There is considerable reason to be optimistic about the transition from income support centers to job centers given that staff there are particularly enthusiastic about helping applicants find jobs."

The decision also called only for a temporary—not a permanent—halt to the job center conversions. The court-ordered pause would have little effect, Turner surmised, on his goal to "reverse the destructive effects of the old welfare system."

Turner made fewer public appearances after the judge's ruling. His plans to revolutionize the entire system by the spring appeared to be perpetually hogtied by the court. Searching for answers, Turner pointed to New York's "cumbersome bureaucratic system" and its reluctance to give up the idea of welfare as an entitlement. "We still sign up people who ask for assistance as we're obligated to do. That's New York State law," Turner said somewhat sadly. "I would prefer the welfare check be made much more like a wage check."

Fazed but not daunted, Turner turned his attention toward the mayor's millennium promise to tackle two new and formidable tasks. One was to parcel out nearly half a billion dollars in federal and state contracts to private and nonprofit job search companies. The contracts were meant to be a key part of Turner's plans to involve the private sector in matching welfare recipients with jobs. Subcontractors would be paid according to how many matches were successfully made. It was a good idea that would prove to be even more fraught with lawsuits and roadblocks than the conversion of job centers.

The second project caught the most public attention. Turner was charged with identifying all those left stuck on the rolls—the system's most vulnerable substance abusers, the mentally ill, the disabled, the homeless—and to find something for them to do in exchange for their checks by the new year.

"People on welfare," vowed Turner, hopefully, "will all be working."

Part III

Hope Deferred

Hope deferred makes the heart sick; a wish come true is a staff of life.

—PROVERBS 13:12

10

Christine

Lost Children

Nine days before Christmas 1998, hallelujahs swelled from the Iglesias de Dios Tabernacle onto 138th Street in the Bronx. Christine conveyed herself and her four children up the concrete steps across the street to the Jackson Avenue Family Shelter. Brenda Fields had made the same climb to temporary refuge two years earlier, feeling much the same relief as Christine did. This was a good day. The Rivera family had spent the last ten weeks crowded into an assessment center in the South Bronx, waiting for a room. Jackson's bright blue and yellow tiles greeted them in the hallway as they scooted their bags down to their rooms. "I am clean. I am almost on my own. This will not be for long," Christine was thinking. "I want to show Luis that I can do this. I know I can do it."

It had taken nearly a year and another temper tantrum at the Crotona Tremont Job Center, but Christine had finally convinced her welfare worker that she was not, in fact, harboring bundles of cash in the bank. She was scheduled to get her first welfare check any day now.

Dyanna skipped into the play yard without looking back. Kristopher clung to his mom. Twelve-year-old Monica and her older brother, Mark, both looked askance at the new place. A cheap motel smell from a caustic cleaning agent stung Mark's nose. Monica announced that she would rather choke than share a room with little sister Dyanna. She had to share anyway. There were only two bedrooms. Mark noticed the curfew rules posted on the door. Another sign said security had the right to check bags at any time. His seventeenth birthday was in a few days. He couldn't have friends over? He couldn't stay out past 10:00 P.M.? Christine searched the notices for stress management classes. She thrived in those kinds of get-togethers, where she could be treated like a functioning adult instead of an inadequate addict. She was determined to graduate from NarcoFreedom, the drug treatment program she was required to attend by the city's child welfare agency. She would get herself to classes at the Cardinal McCloskey center for parents in trouble, another mandatory program. She would beat all this, get her old job back, and bolt to a new apartment, becoming a new woman.

One of Hank Orenstein's staff members checked in Christine and her kids: "Client seems very pleasant. Children were clean. There was no sign of substance abuse or child abuse," wrote case manager Elsie Wood. "Her behavior was pleasant," she repeated.

Christine was asked to write down her future goals: "Hope to be finished with my programs. I would like to go back to work." When asked what could have prevented her from becoming homeless, she wrote: "If I had been working or didn't have a baby, or if I had good child care." Elsie asked her to detail whatever strengths would help her get an apartment: "The fact that I'm doing what I have to do to get my life together."

• • •

It was a tricky time to be homeless in New York City, City shelter was harder to come by than when Brenda Fields was navigating the system. More people were being rejected now. The year Giuliani assumed office, 365 families were denied emergency shelter. By the year Christine applied, that rejected number had ballooned to more than 14,000. A New York survey of homeless shelter providers found that demand for help had increased 90 percent since 1995. The same Coalition for the Homeless report found that three out of four homeless clients were having trouble complying with the new work rules. The vast majority (more than 80 percent) were experiencing "dramatic losses" in benefits, according to a state survey. The rest of the nation's cities weren't faring much better. The Conference of Mayors reported that between 1997 and 1998 requests for shelter rose 15 percent in thirty cities.

If it was hard to convince New York City officials that you were homeless in 1998, it was becoming harder still to keep your room once you got one. The city was putting plans in place that would require homeless residents to work not only for their benefit check but also, indirectly, for their room. For the first time in city welfare history, the new rules would braid public assistance with shelter for homeless clients. Before, if you were homeless and you broke a shelter rule, you could face eviction. But the new plan stipulated that if you were homeless and you broke a welfare work rule, you could lose not only your shelter but your benefits. If you had children, you could lose them too, to foster care.

Christine had more than enough trouble opening her welfare case and keeping it open under the old rules. She missed getting mail, missed making appointments, missed filling out required questionnaires. Now that her children's shelter depended on her keeping track of all these details, she experienced a paralyzing fear of unraveling.

For the mayor and the welfare commissioner, this new homeless-welfare ruling was the next logical extension of the social contract. If a person accepted the gift of shelter, she was expected to give something back for that favor by way of work. It was an important cog in the administration's plan to have every recipient "engaged in work activity" by January 1, 2000—even homeless people. Commissioner Turner believed it was unethical for people to accept free room and board without lifting a finger in return. That was not how the real world operated. The same consequences that applied in the marketplace, he argued, should apply for a room in the shelter.

If a parent was evicted and had nowhere else to go, she would not be permitted to take her children into the streets. In that case, the city would assume responsibility for the children and place them in foster care. Nicholas Scoppetta, the commissioner of the Administration for Children's Services, reasoned that no humane city would stand idly by and watch children freeze in its culverts and on its park benches because their parents couldn't follow the rules. He figured that only the most dysfunctional parent would refuse to work given these odds. "I'd be surprised if any parent who has any sense of responsibility would say, 'I'm not going to take this job,' " Scoppetta testified before a city council committee.

The mayor and his commissioners were surprised by the public pelting they took when these plans were announced. Giuliani's not-yet-official Senate opponent, Hillary Rodham Clinton, said the mayor was treating poverty as a crime. The talk show host Rosie O'Donnell clucked that Giuliani was "out of control." Critics invoked a Gothic vision of Victorian workhouses and a labyrinth of bleak streets inhabited by ragged, tattered families. The city council speaker, Peter Vallone, said the policy conjured up "something out of Charles Dickens." Other welfare advocates imagined another British writer, George Orwell, musing over a plan that first evicted families from their homes, then removed the kids because they were homeless. Somehow, most citizens were willing to accept the armies of street cleaners in WEP programs, but not the specter of homeless families tossed out of their city shelter for disobedience.

Lawyers argued that such evictions were not only irrational but illegal. The new ruling undercut an unusual city ordinance called the "Callahan decree," two decades old, which guaranteed clean and safe shelter to anyone who sought a bed. That order had been passed after the World War II veteran Robert Callahan was turned away from a city shelter into the frigid city streets one winter night because there were no more beds. The Bowery drunk was found face down in a Manhattan gutter. A class action suit was filed after that tragedy, spawning the rule that had become gospel in the shelter system. The legal aid attorney Steve Banks pointed to another 1989 court decision that barred the city from removing children just because their parents had no homes. Invoking both these laws, Banks acquired a temporary injunction to stop the new work rules.

Giuliani argued that the administration's critics were blinded by "fuzzy-headed" liberal ideology. This was a compassionate program, he said, meant to build on people's strengths, not coddle their weaknesses. People would not be booted out willy-nilly—only if they were able-bodied and *refused* to work. The severely mentally ill would be spared, as would those over sixty and mothers with children under three months old. Not spared, his deputy admitted in court papers, would be those in wheelchairs, or others suffering from asymptomatic HIV infection, diabetes, heart trouble, or vascular disease. "The law says you have to put people in a work situation if they're getting welfare, for their good, for their benefit," Giuliani told Nina Bernstein of the *New York Times* in response to news that the vast majority of city shelter directors would unite to defy this rule. "If they're so ideologically opposed to that that they can't carry out the law, then of course they'll lose their city contract."

When Jackson social services director Hank Orenstein heard

about this imminent plan, he figured it was a sign, his personal alba-tross. He could never envision any incarnation of himself as a sort of Simon Legree of the shelter, evicting families who had nowhere to go. Hank believed that most of the people whose cases were closed had not refused to work. They were just overwhelmed by life's chaos and the city's new Rube Goldberg mazes. He also knew that it would be himself, not the mayor, who would be hustling the families and their belongings out the door. No one was really talking about where all these people were going to live in the long run, even if they continued to abide by the rules. The bottom rung of the real estate market was now out of reach for many working poor. The city virtually had stopped building new affordable housing when the Republicans Gov-ernor Pataki and Mayor Giuliani came into office. A new, privately financed apartment complex near Yankee Stadium was able to rent only a small number of its apartments by the time it opened on Feb-ruary 1, 2000. Of its ten-thousand-plus applicants, only a handful earned the $14,000 minimum salary to qualify.

Hank had put in seven good years trying to shore up the broken lives that showed up at the South Bronx shelters he helped run. He was weary of patching up wounded families only to send them out into a world without enough decent jobs or affordable apartments. It seemed counterproductive, and somehow immoral. He badly needed a career change. The Fordham University lecturer was planning a per-sonal move from Queens to Manhattan, with his wife and kindergarten-age son. His work in photography had already evolved from expansive vistas of lonely valleys and canyons into minuscule landscape abstrac-tions, created almost by accident with paint drops and organic mat-ter on a microscope slide. In his working life he wanted to move in the opposite direction, from the micro to the macro, from directing the shelter to addressing the larger reasons for its existence. Hank accepted a position as director of child planning and research with the city's public advocate. There he could help shape policy, not just react to it. So after several tearful farewell parties, Hank left the Jackson Avenue Family Shelter.

A month later Manhattan Supreme Court Judge Stanley Sklar ruled that the mayor's new homeless-welfare plans violated the homeless person's right to shelter. He was most moved by mental health experts who testified that all the new workfare rules made it much more difficult for the most fragile people to avoid breaking them. The result, he predicted, would be "an explosion of homeless individuals, banished or barred from shelters, risking their health, and perhaps their lives, on the often bitterly cold and palpably dangerous streets of a sadly indifferent city." For a moment it seemed as if the nineteenth-century muckraker Jacob Riis had returned to write court briefs.

The mayor responded that the judge was "clinging to his desire for a city of dependency." His attorneys filed an appeal.

Before the ruling came down, Hank turned over the reins of the Jackson shelter to his younger, energetic protégé. Joseph Esheyigba, a coal-black immigrant from Nigeria with a melodious baritone, had studied social policy under Hank at Fordham University. Shelter residents stumbled over his last name and called him simply "Mr. Joseph." He returned their respect by working sixty-hour weeks sorting out their lives, counseling them in a voice marinated in the rhythms of his African homeland.

In many ways beyond race, Joseph was Hank's opposite. Hank preferred casual shoes and loose corduroys. Joseph came to work on East 138th Street in pressed suits and starched white shirts. Raised in a fairly conventional Manhattan home, Hank oversaw the roiling South Bronx shelter with a calm detachment. He responded to people as he did his art—with a quiet respect for the complexity underneath. Joseph, raised as a virtual orphan in abject African poverty, charged through his day with an emotional zeal for this often exasperating work. His clients knew when he cared, and they knew when he'd reached his wit's end.

Joseph's parents left him with his grandmother in Lagos when he was two years old so that they could attend college and earn a better living in London. He and his younger brother barely saw either parent again. The boys were raised in an apartment with no running water, above a shop in a seedy market area. Joseph and his younger brother used a pot for a toilet. They slept three to a bed. Joseph's first after-school job was cleaning up after the prostitutes at his stepgrandfather's Lagos hotel.

Friends and relatives recognized Joseph's bright potential and pooled ten dollars here, twenty-five dollars there, to pay eventually for his passage to London. The staid British Isles never suited the spirited young African, so he stole away on a tourist visa to New York City as soon as he could. He worked long hours at a tile factory in Hunt's Point for years until immigration officials nearly caught up with him. When he finally obtained his working papers, a more focused Joseph took a job as a security guard at the New York Telephone office on the Grand Concourse during the day, while studying for his community college degree at Baruch at night. One day a child welfare caseworker spotted Joseph studying at his security post and suggested he might consider working in child protection services. That serendipitous encounter set Joseph on a career path serving first crime victims and later the homeless. Now married, with two daughters, Joseph was still living an ambitious, frenetic existence, supervising caseworkers during the day and studying for his master's and later his doctorate in social work at night at Yeshiva University.

Joseph was no more pleased with the new work rules than Hank. He would fight them on the frontlines. Hank would tackle them on the policy side.

• • •

Joseph watched one summer day as school-age children ran back and forth between the Cypress Avenue subway stop and the front steps of the Jackson shelter for some presupper playtime. A shirtless man, his forearms a painted swirl of tattoos, wandered past with a boa constrictor wrapped around his neck. The kids suspended their race to skip along beside him, thrilled by the slow, slithery creature. Other enterprising children played basketball behind the shelter using a plastic milk crate with the bottom carved out. The plywood backboard was painted with a Puerto Rican flag. An old mattress was wrapped around the streetlight-slash-basketball pole so that the boys could drive to the crate without fear of concussion.

"There's not much to do for the older children here," Joseph said. "Shelter rules make it hard for them to play inside. This is not a good place for them." He was intent on getting the children and their parents out of this particularly bankrupt corner of the Bronx. Although shops and chain stores did a brisk business along the newly revitalized 138th Street, drug peddlers still sold their goods every day around the corner on Cypress Avenue. Prostitutes walked the streets just west beyond the Bruckner Boulevard overpass. Joseph's pet project was to organize trips for the kids to the Coney Island aquarium, or to a Yankees game, traveling from the East Bronx to the opposite side of the borough.

"Christine needs to get out of here. She needs to find an apartment far, far away from here. Too many temptations," Joseph said. He was fond of Christine. With everything she had been through, she was still trusting, still thoughtful. She sometimes dropped by his office to tell him how her apartment search was going, or how a welfare hearing went. Not many residents would think to do that. Joseph noted that she loved her children and was receptive to those who wanted to help her. The trick with Christine was to remain nonjudgmental. "If you don't get confrontational," Joseph said, "she is delightful."

Christine tried her best to fit into shelter life. She was grateful for the free Pampers and the vouchers for cheese and milk until her food stamps arrived. "I'll get my apartment. I'll get my job back. I'll show Luis I can," she said almost every day. But within a few short weeks Christine felt like a Coney Island bumper car, colliding with the rules and regulations and with their chief enforcer, Elsie Wood, the shelter social worker who had greeted her with hope that first day.

Elsie was an old-school caseworker who in many ways embraced the sentiment behind the new welfare-to-work laws. She believed that character flaws might just be at the root of her clients' troubled lives. She had little patience for much of the immature behavior she witnessed. Working for their benefits might help, she thought. It might be the best way to structure these residents' chaotic lives. The kinds of people who ended up in the shelter needed more organization, and they needed to learn how to be better role models for their children. She thought most of them could use a firm hand. Elsie befriended many of the shelter security guards, who often called her at home to report the residents' various misdeeds. Christine had slammed up against many Elsies in her life, and she never did learn how to get along with them without a blowup.

Two blots on Christine's child welfare record didn't help her image as a functioning mother in Elsie's eyes. The first was that Kristopher was born with opiates in his system. The second happened right before Thanksgiving 1998, in the Powers Assessment Center. Christine sent both her girls off to school on the city bus one morning. Monica, the junior high student, was supposed to escort Dyanna, her younger sister, on the mile-long ride across the tip of the Bronx to P.S. 40. But Monica was late, so she put Dyanna on her own bus and headed off north to her intermediate school in Co-Op City by herself. The bus driver found the little girl curled up with her book bag, asleep, at the back of his bus when he finished his route. He eventually taxied Dyanna to her elementary school. The incident got back to

Christine's child welfare caseworker. "That was poor judgment," Christine admitted. "I shouldn't have done that."

With Elsie and the security guards watching her every move, Christine felt she was back in Holy Cross Catholic School, where the nuns gazed down disapprovingly upon her weekly dollar in the offering plate. ("They told my mom we should give five dollars. That really made me mad. We didn't have it.") Their constant disapproval made her crazy.

The second day at Jackson, Elsie reprimanded Christine for having a messy room. The third day she lectured her for letting Monica and Dyanna play in the hallway by themselves. A week later Elsie told the harried mother to put her children to bed by 10:00 P.M. On another evening, Monica sneaked out without her mother knowing and came back after curfew. Security caught her. Then Kristopher, the baby, toddled down the hall away from his mom, who was talking with a friend in her room. Elsie scooped him up before he fell down some stairs, then castigated the errant mother again.

All infractions ended up in a handwritten pile called "Homeless Programs Incident Reports." Under "type of report," the accuser checked "curfew violation," twice, and "breach of rules and regulations" nearly a dozen times. By the end of Christine's first month, Elsie wrote, "Ms. Rivera is trying to make improvements. But after being the way she is for so long, change will be hard. However, the case manager will enable Ms. Rivera to accomplish her goals."

Meanwhile, Christine was still drug-free but practically penniless. Her official welfare budget incorrectly failed to account for all her children. Her benefit check tallied up to $222 per month; food stamps provided $340. HRA normally provided four times that amount for a family of five, plus $400 in food stamps. Even that was barely adequate.

The welfare grant had not changed since 1990, despite cost-of-living increases. It had lost about one-quarter of its value in a decade, and over half its value since 1975. For a typical family, the basic allowance came to $3.19 per person, per day. Food stamps added another $6.25, or a little over $2 per meal. The rent allowance was $286 in a city where the fair market rent was more than three times as much. Many people ran out of food stamps by week three, resorting to public food pantries in order to make it to the end of the month. It was not uncommon for welfare mothers to borrow from others or work in the underground economy as baby-sitters, unlicensed vendors, or exotic dancers to make ends meet. The most desperate, or the least resourceful, turned to stealing or selling drugs.

One of Elsie's responsibilities was to make sure her clients' public assistance cases were up-to-date—a requirement for shelter. Yet Christine's case floundered inexplicably for three months. Finally, in early March, the city closed it down, without warning. Christine found this out the hard way, after navigating Kristopher in his stroller to the check-cashing storefront on the main thoroughfare to collect her check. The clerk behind the protective bars told her she no longer had a case. Christine, always fragile, was dangerously close to her breaking point.

A few days later Elsie wrote in her case notes: "Ms. Rivera was not in a good mood." That was probably an understatement. She had been kicked off welfare for a paperwork snafu. A fraud investigator discovered that Christine hadn't returned some required questionnaire. This was news to Christine. She got her hands on the form, turned it in, and applied for a fair hearing to receive retroactive benefits. For another month she and her family had no money whatsoever. Joseph would loan her subway fare so she could go to her rehab program and search for apartments. Elsie would give her vouchers for the food pantry, Abundant Life, so she could keep her kids in peanut butter and cereal. Luis was strapped himself but always came up with at least twenty dollars a week to give to her. Eventually Christine won her hearing. More than 80 percent of the complainants won their cases during the Giuliani years—a strong indication that most welfare sanctions were not warranted. "What does that tell you?" Joseph asked in exasperation. "That tells you people are getting sanctioned for no good reason. I see this all the time."

Christine's typical day was a blur of random required activities. She got her girls off to school, then took the subway to Brooklyn to straighten out her welfare mess. Back up to the Bronx for rehab and a parenting program that so far had exempted her from WEP. Then over to the shelter to pick up Kristopher from day care, unless he was sick, in which case he would come along with her. There were meetings with caseworkers, more meetings with housing experts. A letter would arrive telling her to report for a WEP job or lose her benefits. She would schedule and reschedule appointments to find out what to do. Some NarcoFreedom women were being pulled out of classes to clean municipal buildings and sweep the parks for their welfare checks. How would she ever fit that in? What good would it do her? Another letter would come saying she had one day to apply for federal Section 8 housing. She had to beg for subway fare to make all these appointments. If she missed an appointment, she was in danger of losing her welfare, and later her shelter and her kids.

Everywhere Christine went she was late. Everywhere she went she got yelled at for being late. She hated being scolded, attacked. She hated that foxhole feeling.

Christine's old depression began to grip her days and nights. She tried to fight it off, without much help. She would attend little ceremonies at NarcoFreedom where other women received plaques and applause for completing their steps. Christine really wanted a plaque, she confessed. She wanted to finish something. She was desperate for that feeling of accomplishment. She wanted an audience to clap for her. She'd had it once before, when she passed a data processing program, several years after her mom was killed in the tragic fall from a fifth-story window. No one thought she could do it then. It was the best she ever felt about herself.

Then word trickled down to Christine's second-floor room that another Jackson resident was dealing drugs. Before he was caught and thrown out, Christine became a client, using heroin again, on and off. There was no excuse for this, she knew, but the addiction was stronger than her depleted willpower. She also had a ready-made rationalization. Whenever she had worked hard to stay clean, the workers would accuse her of using anyway, so she decided: What's the use?

Within a short time she was hooked again. She skipped some NarcoFreedom classes—sometimes because she had dirty urine, more often because she had conflicting appointments. She skipped parenting classes, sometimes because she had dirty urine, more often because her baby was sick. Kristopher had asthma. The South Bronx suffered the highest rate of childhood asthma in the nation. One of the main reasons poor mothers had trouble keeping steady employment was because their children required constant hospitalization. Christine wasn't allowed to bring Kristopher to Jackson's day care when he was sick. Now she had drug problems besides. She'd succumbed to the environment's poison, again.

Finally, Christine agreed that the only way out of this downward spiral was to go to a detox program in the woods near Allentown, Pennsylvania. Her daughters stayed with a distant friend on Long Island. Her sons stayed with Luis. The program sounded like a fantasy to her—White Deer Run. She envisioned deer running free, under the sky of black, crisp nights. She would feel like she did on the beaches near San Juan with Luis. She wasn't disappointed either. Graceful deer ambled right up to the porch to nibble on the bushes. The nights were clear, the stars so close the rays almost hurt her eyes. It reminded her of the air in Puerto Rico's mountains. The staff served luxurious meals: steak, lobster, all-you-could-eat breakfasts. It was a stillness, a satisfaction, she had never known.

The group sessions helped the sociable Christine. One in particular had a grim impact. The instructor asked everyone to name all the people they knew who had died from drugs. She and the other patients wrote them down in magic marker on white poster paper, creating a "wall," as the instructor called it. Soon the wall was a black smear of casualties. All the patients were in tears, remembering their friends' deaths, fearing for their own. They lit candles and sang songs.

Christine wanted to stay for the full month, but she couldn't leave her kids that long. She returned to the Bronx after two weeks. Clean. Renewed. Peppy, and a little chubbier, with all that food. Her long hair was trimmed and dyed with a burgundy tint. She returned to the Jackson shelter, to her relieved children, to the South Bronx, to the watchful eye of Elsie Wood. But two bucolic weeks were not enough to beat a lifetime of street survival and self-abuse.

Something was brewing, something big and probably bad. Christine could sense it. Elsie Wood kept writing her up for breaching this rule or that regulation. Then she quit her job at Jackson. The day Elsie left for another social work position, applause erupted simultaneously throughout the shelter. Other residents, it seemed, also had rubbed tempers with the case manager. Once someone was angry enough to hurl a rock through her office window. With Elsie gone, Christine's stress began to subside temporarily. The new worker, Wanda Abeyllez, helped straighten out her welfare case. She got Christine certified for federal Section 8 housing. Christine could see herself in a new apartment, drug-free. She would show Luis she could do it.

"I felt sorry for Christine," said Wanda. "There was all kind of crazy stuff in her case, and nothing was done about it. I worked for three weeks to get her welfare case reopened. There she was going around in all the weather, with the baby, making these appointments."

A few weeks after Mother's Day 1999, Christine took Kristopher and Dyanna outside to 138th Street after supper to talk to friends. A rhododendron curled up from thirst in the gated garden out front. Tempers and baby cries competed in the thick city night with the rap and reggae music streaming from open windows. Christine had just seen an apartment that afternoon—a walk-through near Lincoln Hospital. There was a church on either end of the street, like bookends. It felt right. She was telling her friends about it.

After a while Dyanna got bored and ran back inside, from the sidewalk up the stairs and into the shelter. "I knew where she was," Christine said. "I knew she was okay." When Christine returned to the shelter an hour later, the security guard said she had left her daughter alone and asleep in the visiting room. He reported her. It wasn't Christine's most serious offense. Still, it turned out to be her last—the final consequence for ending up in this South Bronx fishbowl. Joseph called child welfare.

The next morning the Administration for Children's Services caseworkers came to Dyanna's school to take her into their custody. ACS workers ordered Monica to come to their Bronx office when she returned from junior high. Salimie Calim, Christine's main child welfare caseworker, came to Jackson herself to take Kristopher, now two years old. Christine was oddly grateful for Salimie's gentleness. She didn't take kindness for granted. "She could have come with police, and she didn't. That scares the kids more," Christine said. Christine put her son in Salimie's car, strapping him in the car seat. "I was crying and crying. Kristopher saw me crying, and he got tears in his eyes. It was almost like he knew what was going on. He couldn't know. But he seemed to. I was a mess."

Salimie hated that part of her job. She dispatched these often unpleasant duties with a sense of sad inevitability. "I had no choice," Salimie said, with a distracted shrug. She mentioned the bus incident, the dirty urines, the chronically messy room, the unsupervised children. Eleven rules breached and two curfew violations added up, in the final incident report, to "child abuse with removal of children, but no arrest."

Christine had children around her for half her life. Now they were all gone. The boys were with Luis. The girls were in foster care. Dyanna, seven, would be all right, she thought, as long as the adults were kind to her, and as long as she had her dolls. Monica, twelve, was more of a worry. She was angry. She had a temper. For now, this room in the shelter felt hollow, the empty nest of a mother bird with failed wings. There was no one fighting over anything. No one grabbing on to her leg. There was no one to scream at: "Pick this up, pick that up." Just silence. She felt unhinged.

Alina

Reluctant Refugee

One month after Christine lost her grip and her children, Alina tiptoed into her journey back in time, to discover her future. What exactly had she fled from? Was it a mistake for her to exchange the shelter of her homeland for an uncertain future on America's welfare lines? She was anxious to at least ask the questions, though unsure if she was prepared for the answers.

A flight from JFK Airport in New York carried Alina to Amsterdam, and to a less-traveled connection. On a cool day in June, she entered through the rear of the Air Moldova jetliner in Amsterdam. At the top of the narrow stairs under the tail was a prim stewardess (not a flight attendant) holding a tray of flat Cokes in small glasses. Her smile was all lipstick, red and practiced. Her outfit was early airline: plaid vest, tight skirt, high heels, and shellacked hair. No strand dared to stray. No passenger went unattended. The year felt more like 1959 than 1999.

This Yak-40 was one of a handful of aging Soviet jets in the airline's fledgling fleet, reconstructed and repainted to fool no one. The word "stewardess" in English and Russian was repainted and emboldened in black next to a button above everyone's seat. Moldovan, the official language, appeared only in the glossy in-flight magazine that introduced readers to the airline's new twenty-six-year-old general director, who was quoted as saying he enjoyed billiards and "dance disco."

Alina tossed her bag onto the open shelf above the seats. This was a Greyhound bus–style rack, where an inch-high rope separated bulging luggage from the passengers' heads. Others placed their baggage wherever they liked, in the aisles, on an empty seat. Passengers wandered the cabin during takeoff and landing, paying little attention to the seat belt pings or the tray-table regulations. Moldovan travelers either were untrained in the conventions of plane decorum or just didn't care that their bags could bop a fellow traveler. Theirs was a country with only a fleeting experience in the art of the lawsuit.

Stewardesses flitted from one seat to the next during the flight's duration, offering a nonstop cornucopia of jelly-filled candies, cold cuts, sausages, *pastiki* meat spread, sour pumpernickel slices, plum cakes, sodas, tea. It was a deceptive display of bounty that was far scarcer in the country that awaited. Alina nearly swooned from the familiar smells and tastes. She stuffed the bits she couldn't finish in her pockets for later.

Alina was making this trip as a young woman without a real country. She was neither an American citizen nor a Moldovan. Her red Moldovan passport issued by Soviet officials was not recognized in the new republic. The country's embassy in Washington, D.C., had warned her that officials would confiscate it at the Chisinau airport ("Chisinau" is Moldovan for the Russian "Kishinev"). So she applied instead for a refugee travel document from the U.S. Immigration and Naturalization Service. Alina's new identity became, temporarily, a U.S. refugee with a visa for Moldova. Her papers never let her forget her tentative position in the world as a displaced person.

It was somewhat of a surprise that Alina agreed to make the trip at all, at the urging of this author, through a travel grant from the Freedom Forum Professors' Publishing Program. Her mother did her best to convince her daughter that Kishinev muggers preyed on anyone overheard speaking English, or even Russian. Her fiancé, Mark Filizov, a business student at Fordham University, worried that she would never come back home. He knew that Alina was not as taken as he was with New York's fast pace and fast money. Her real sense of well-being still lay somewhere in the rattling trolley cars and street dances of Moldova's capital. As for Alina, she worried about having enough of the right kind of food to eat. She had no experience with Kishinev's hotels or restaurants. Under Soviet rule, tourist areas were off-limits to citizens. It wasn't clear whether all this anxiety was more of the infamous Zukina nerves or simply her overactive common sense. In the end the pull of her homeland prevailed. She felt disconnected from herself in America. She had left two dear friends in Kishinev, both named Natasha, to whom she never properly said farewell.

Back in New York, Alina had packed her suitcase with sausages and dried soups—enough for a week—some U.S. dollars, and thrift-shop sweaters and shirts, which were gifts for the few family friends left in Moldova. She figured her return would be cathartic. The idea that she could explore her American life and her Jewish identity as a mature woman with relative freedom had some appeal. She might make better sense of her last four years as a reluctant refugee who was grappling with poverty, a hostile bureaucracy, and a future as a hyphenated American. Perhaps she would finally put to rest the gnawing sense that she'd made a fundamental mistake in ever leaving the country she loved.

The Yak-40 landed without incident in what looked like a bucolic Balkan outpost. Old rolling hills, some of them harvested bare, some planted with low grape vines and clumps of fruit trees, surrounded the Kishinev airport. A makeshift trailer marked "International Airport" worked as an unintended decoy for the long concrete terminal right next to it, which was under construction. A thresher stood idle by the runway. Passengers, many still without the necessary customs documents, entered a plywood shed containing a rickety conveyer belt for luggage. A mob of visitors and cab drivers hollered "Taxi! Taxi!" and shouted the names of passengers from an open door a few yards away in the airport lobby.

Alina noticed her friend, Natasha Isakova, waving frantically above the hot crowd. The raven-haired Russian, dressed in her finest shiny gold skirt and top, whisked Alina out of the crush of people and into the parking lot, where they climbed into Natasha's minivan. It was rare for a twenty-three-year-old to own a vehicle. With an inheritance from her grandmother's small estate, Natasha had purchased a government-owned apartment and stocked it with new, Romanian-made appliances. Still, her eight-dollar-per-month nursing salary made it barely possible to keep up with utilities and taxes, let alone to buy food and clothes. The made-in-Russia van brought her a little extra income. She hired a young driver, Sergei, her on-and-off boyfriend, to work one of the many private transportation routes that sprouted up after independence. The van owners charged about eight cents per passenger to ride unregulated around unmarked routes in the capital city.

Natasha seemed to be in a huge hurry to leave the airport, but her funky minivan with little orange curtains didn't cooperate. Sergei climbed under the engine while Natasha worked the gears, each in turn yelling Russian commands to each other. After several grinding tries, they were finally off, heading north to Moldova's unassuming capital. The van passed between the city's two trademark white apartment buildings that formed a topless welcoming arc. Close up, many of the buildings' windows were missing, giving the once-grand gateway a toothless look. The main boulevard was lined with blocky Soviet structures, occasionally interrupted by Turkish-style municipal buildings in baroque pastels—reminders that this city had been almost

completely obliterated in recent history by Nazi air raids and an earthquake thirty years later.

Under Soviet rule, the tree-lined promenade was called Lenin
Street. The Moldovan republic had since renamed it after Stefan cel
Mare, or Stefan the Great, a fairly obscure fifteenth-century prince
known for forging an independent Moldavian state in the midst of the
Ottoman empire. Robert Kaplan wrote in *Balkan Ghosts* that Stefan
the Great built brightly painted monasteries in order to spread the
word about Christianity and Moldavian culture to the peasants, defying the Turkish oppressors. For these reasons, the little king (*nu mare
du stat,* "not great by size" at only five-foot-four) became the ideological symbol of the new Moldova.

Few protesters knew of this history when hundreds of thousands
gathered at the feet of Stefan's statue in Pushkin Park during the early
nineties to stir up nationalist fervor for Moldova's independence from
the Soviet Union. More Romanian than Russian, the majority
Moldovan population grumbled under what they felt had been
decades of arrogant Soviet domination. For the native-born, independence had been a half-century in coming. At first Alina and her
friends had found this new interest in the little nobleman amusing.
But when the crowds became larger and more intense, so did the
xenophobic sloganeering. Signs calling for the new republic to
"Cleanse Moldova of Jews" frayed the edges of the country's already
fragile social fabric. Riots threatened, but never materialized. At one
point a protester broke from the ongoing rally and rushed Alina and
Mark, who were walking hand in hand through the park. The
Moldovan man yanked their hands apart to provoke a fight with the
Jewish couple. Alina pulled Mark away before an incident could erupt.
As for Natasha, she never came near the rallies. She was pure Russian,
as was one-quarter of Kishinev's population of 600,000, and thus a
modern-day equivalent of the ancient Turkish occupiers.

Sergei pulled the van up to Natasha's apartment building, part of a concrete complex of Soviet-built homes ubiquitous in the outskirts of the capital. These utilitarian concrete blocks, cracked and neglected though only one decade old, were reminiscent of the South Bronx in the late sixties, when landlords abandoned blocks of old buildings to decay from the inside out. The difference was that most Kishinev citizens—engineers, laborers, nurses, shoemakers, cooks, teachers—lived in such buildings. This was not a ghetto separating the poor from the affluent. Here Alina began to realize that while poverty was perhaps pervasive, it was also somehow less isolating than in America. She realized that much of her current willpower came from growing up with a strong sense of self amid equal opportunity deprivation.

Sergei navigated the van around the concrete slabs that once formed a real road in front of Natasha's five-story building. He drove through groups of children playing with sticks on patches of grass and parked right at the front door, which was a blackened hole. A waft of chocolate aroma sweetened the air from the Bucuria chocolate factory a few miles away. Swarms of flies, attracted to the garbage crate by the door, greeted all visitors in the hot June air. The elevator didn't work, so the group walked up the three flights of stairs, past broken metal mailboxes on each landing. This was why Alina rarely wrote letters to her friends.

Inside, Natasha's apartment was neat and roomy. A black wall unit featured a new Sony television, with cable, bought with her grandmother's inheritance. A velour pullout couch decorated the living room. When *CNN News* was not broadcasting the final days of America's NATO bombing campaign against Slobodan Milosovic's Yugoslavia, the other English-language stations were playing bottom-of-the-barrel American movies such as *Hell in Frogtown* or *Revenge of the Killer Tomatoes*.

Since the apartment had no hot water, Natasha washed dishes with cold water and heated pots of water for sponge baths. Every night the electricity was shut off at 11:00 P.M., making it impossible to pump any water through the pipes. The toilet wouldn't flush, the sink basin came up dry. Natasha, with a medical degree in nursing, made six dollars a month more than the average Moldovan citizen, and about seven times less than the monthly grocery costs. Her relatively stable earnings did not immunize her from a total utility shutdown. If tenants on her line didn't pay their personal utility bill on time, the entire block of apartments would be cut off. A few weeks earlier the whole of Kishinev was without electricity for three days running. In the months to come the city would lose electricity three to four more times. Ukraine and Romania would cut off the capital's supply in order to collect the nearly $200 million Moldova owed them both in back utility bills. Romania, Albania, and Moldova all vied for the position as Eastern Europe's most destitute.

"Socialism was better than this. This is a mess," scoffed Natasha with a sarcastic pucker. She called herself an old-style Communist with a healthy sense of distrust of all things Western. "Maybe there are more things to buy now, but there is less money to buy them with." Natasha did own a pair of fake Adidas sneakers and a bootleg Nike sports jacket, but she remained relatively cool about collecting American goods, unlike many of her madly materialistic countrymen. She still longed for the strict Soviet days when workers lounging in front of the cinema during work hours would be rounded up and arrested. "People are getting away with too much," she said.

Alina made a meal of fried potatoes, boiled eggs, brown bread, and a freshly picked tomato from the market. It smelled and tasted like home. "This is what we crave in New York," Alina said. "Fresh. Close to the earth." Natasha shrugged.

The next day Alina began her nostalgic tour of Kishinev to make sense of the present by remembering the recent past. She passed the street where she had said good-bye to Mark at a street dance the night before he emigrated to America before her. She walked by the Frigushor (Snowflake) Café, where she would stop for soft ice cream on

her way home from school, and rounded the corner where her childhood home still stood. Alina gasped when she saw the ramshackle
stone cottage encased in a rust-colored wall. The home was a quaint
survivor from the days when Kishinev was a provincial Romanian village. Nothing much had changed over the century except the tenants.
Several towering balsam poplars had showered clumps of white seed-
filled balls called *pukh* along the street, leaving a puckery wintry coat-
ing on the hot summer road. In other parts of the city, gypsies
gathered the balls in large blankets—but not here.

This was the road near downtown Kishinev where Alina used to
skate on her clamp-on roller skates with a key. Seeing this street again
reminded Alina of the Soviet parades that would routinely emerge
from the park above and march down the skateable hill past her
house. When she heard the drums, she would run outside in her heavy
bearskin coat, so long it picked up a permanent fringe of snow. She
would admire the marchers, imagining someday leading such parades
as a political leader, and later mayor.

She walked up to the wall around her old, disheveled courtyard
and ran her fingers over the large crack that ran from the street to the
top. Her mother frequently recounted the night this stucco scar had
first appeared. An earthquake shook the entire city. The wall next to
Alina's crib collapsed. Because of that event, the Zukinas still shuddered when an overhead subway train shook their building in the
Bronx, remembering that day their house had cracked open twenty
years earlier. Looking at the place now, Alina was surprised the little
fragile cottage hadn't crumbled into dust.

A passerby in a rayon business suit stopped to puzzle over why an
American would be taking a picture of this ancient-looking slab. "Take
pictures of the new Moldova," she encouraged. "Not these old things."
Alina didn't mind the patriotic insult. She appreciated the home's
earthy charm more than her mother complained about its primitive
lack of comforts. Inside the medieval wooden door was a bare yard

lined with two sheds and an open stone cellar. A terrier yapped under the clothesline at intruders. A rooster and two hens squawked and clucked around a picnic table. Alina remembered most of all being cold in this house, in the middle of the capital city. The Zukinas heated the rooms by stoking a stove with coal or wood, but the fuel was never enough. "The ink in my pen would freeze up and not work some-times,"said Alina, who lived in this cottage until the age of twelve. "Sometimes I couldn't finish my homework." The drafty walls kept her in sniffles, a common affliction her mother tried to remedy with vari-ous undrinkable concoctions of raw eggs and honey, or raw eggs and baking soda.

"I would often criticize my mother for wasting wood or coal, or using too much water from the fountain, things I learned in school," Alina said. A lot went unsaid between mother and daughter, such as why Lyusya shushed Alina from talking about politics or religion while standing on public lines, or why Lyusya never taught her daughter Jewish rituals. Now Alina realized her mother was protecting her from government spies or other forms of discrimination. "I believed every-thing I was taught in school, that everything Russian was better than anything else. I thought the best way to get ahead in Kishinev would be as a political leader in the Communist Party. I would organize park cleanups. I would work for the party."

Moshe was first brought back to this cottage as a fragile infant, when Alina was eleven. Their thirty-five-year-old mother had been ill throughout the pregnancy, suffering from extremely high blood pres-sure. Lyusya bribed a doctor to make sure he would attend Moshe's birth at the hospital maternity ward—standard practice in the old Moldova and the new. Without a bribe, there would probably be no attending physician. But something went wrong. Even though the family paid him the required rubles in advance, the doctor was nowhere to be found when Lyusya went into labor. She gave birth with an inexperienced aide in attendance. Moshe's collarbone was broken during the rough delivery. As a result, mother and son were forced to stay for a long time in the pale-yellow maternity ward, separate from the general hospital. Alina remembered standing with her father in the grounds outside the baroque building, looking at her mother holding baby Moshe in a window framed with white gingerbread trim. Visitors weren't allowed inside. An iron fence, its spokes poking in eerie, off-angle directions, surrounded the overgrown grounds, which were thought to be the site of an old Jewish ghetto and cemetery. When staph infections ran rampant through the baby hospital, patients blamed angry Jewish ghosts for stirring up disease from their brambled graves.

Moshe required a long time to walk. He first crawled on his fingertips only. "Not a good sign," Alina explained. "My mother was so worried, she took him to a gypsy when he was two years old to read his fortune. The old woman took an egg in her fingers and shook it all around Moshe's head. Then she broke the egg in a glass, and dropped candlewax in it. Then she read the wax droppings. She said Moshe would walk when we returned home, and he did. She also said I would marry a doctor, and study real estate." Neither of the gypsy's predictions for Alina had come to pass so far.

Being Jewish was an embarrassment to Alina, an unspoken badge of otherness she didn't quite understand. Alina remembered blushing hot red when a Soviet teacher once told the class that many Ashkenazi Jews were mentally slow because they had a long history of intramarriage. She imagined her classmates' eyes boring through her, searching for signs of genetic stupidity. Still, she had no reason to doubt the teacher, no compendium of research to rebut her claim, even if she wished.

Attending synagogue was as rare an event for the Moldovan Jew as was lighting candles at home on the Sabbath. It was not only taboo, it was dangerous. Alina's father sometimes braved the hail of stones that would often rain on templegoers in the old Jewish quarter for special occasions—a high holiday or the funeral of a friend or relative. On Passover her father would return from the synagogue with a bag full of matzohs, the ritual flat bread used for Seder. "We would eat it for the rest of the year with our meals, not really understanding what it meant," Alina said. "My mother didn't let us work on Saturdays. But I never knew why. We never talked about it. I never knew what is kosher, or what is Yom Kippur." Her mother hoped a shroud of religious ignorance would shield her children from discrimination.

Alina's spiritual world began to uncoil after perestroika. The Jewish Library opened across the street from her home, with a Holocaust display of memorabilia and a vast collection of books. The first Jewish school opened up in Kishinev in 1991, funded jointly by the Israeli and Moldovan governments, with help from Jewish philanthropies. At age sixteen, Alina transferred to the new facility and became one of fifteen students in its first graduating class. The secular Jewish School No. 22 taught Alina to be proud instead of afraid of her heritage. "It was sad, but I went into a synagogue for the first time in my life that year," Alina said.

The high school student was most impressed by Jewish contributions to modern science. "We learned about Einstein before, but I never knew he was Jewish." Funds for the start-up school were so scarce that students were called upon to do the work of janitors. Children washed windows, cleaned the rooms, built furniture, and found books to stock the library. "No one complained because we were all glad to be there," said Alina. "I felt comfortable in that school. It was the first time I felt that." The new hopeful sentiments were not great enough, however, to withstand the barreling economic catastrophe. Nearly every student in that first class, including Alina and her boyfriend Mark, and most of the teachers fled Kishinev with few resources. Despite Moldova's new veneer of religious freedom, it still offered a dead-end future economically.

The exodus of Jews had continued every year hence. Officials estimated that forty-four thousand Jews left Moldova after 1989—a few thousand more than the number who remained. The majority settled in Israel. The second choice was America—a country Alina knew only as home to socialist author Jack London, and a host of cruel eviction policies toward its poor. Alina's path was well traveled, and her experience with deprivation well lived.

12

Christine

Invisible

Christine's journey, by contrast, would be a lonely one from here on. Luis Lopez arrived by himself at Bronx Family Court early on June 2, 1999, the morning a judge would decide what happened next to Christine and her children. Luis moved through the magnetometer line in his dark blue suit and white shirt with the air of authority he brought from his years working in criminal court around the corner as a Spanish interpreter. The new foster mother, a grandmotherly Dominican woman, arrived soon after, escorting the girls. She'd been caring for them for one week so far. Monica stood sad and tense, her hair pulled into a neat bun, her white shirt pressed and starched. She didn't know what to expect next in her uprooted life. Dyanna wriggled in line, wearing a pink jumper and colorful barrettes, alternately whispering in Luis's ear and playing hand games with Monica.

The girls seemed worried, but content with their new caregiver. Monica had sneaked a call to her mom a few days earlier, reassuring

Christine that she and Dyanna were fine. The woman didn't speak any English, but was nice. She bought the girls new clothes with the four hundred dollars per child she received from the city. She was an emergency caregiver, the first stop for neglected children before a permanent home could be found. She ran a cake-baking business in her home on the side, and promised Monica a magnificent pineapple and cream concoction for her upcoming thirteenth birthday.

By night, police believed, rooftop sharpshooters used the Bronx Family Court building for target practice, judging from the bullet nicks on its concrete facade. By day the building up 161st Street from Yankee Stadium served as a figurative battlefield: nearly fifteen hundred of the borough's poorest families arrived here every day to wage the most intimate combat of their lives. At this early summer hour, the court crackled with hot anxiety disguised, for the most part, as resignation. Each person in the security line had experienced some overwhelming personal crisis that led him or her here. Now teams of lawyers, social workers, and judges were in a position to reach into the already chaotic lives of their families and alter them forever.

Some litigants were dressed in their Sunday best, the women in light cotton dresses and high heels, men in ties and dress pants. Others came more informally, in jeans and sneakers, in rayon sport jerseys and mini-hoop earrings. Some were in the court to defend themselves against charges of abuse. Several children, under sixteen, faced allegations of assault or theft. Young men, swaggering in saggy jeans and sleeveless T-shirts, took off their belts in anticipation of the approaching metal detectors. With one hand, they clutched packets of documents bound by rubber bands. With the other, they held up their pants.

Luis was in court to win temporary custody of the baby, Kristopher, who was asleep in the stroller. At first the upstairs waiting areas were serene. Luis looked for the handwritten sign identifying the right judge's name, Gayle Roberts. Plaintiffs had left messages carved into the wooden waiting benches: "Harry 4Ever," "Yo quiero Alberto." The day's cases began to filter into the room. Two designer-clad teenagers arrived. The girl carried an infant strapped to her torso papoose-like. The boy was armed with a backpack full of papers. They cut menacing glances at each other across the room.

On another bench a little girl in a Winnie the Pooh baseball cap recited the months of the year in English and Spanish for her beaming mother. A few benches away, a middle-aged white man rocked in place, clutching a ragged manila folder. His ears were crusted white, his thin hair flew in all directions. His race drew attention as much as his incoherent mumbling. Most everyone else was black or Hispanic. Once, the man stood up and approached the little girl in the Pooh cap. The girl panicked, stretching her arms in front of her to push him away. Conversations stopped cold in the room. He told the little girl he wouldn't see her for twenty years. He was leaving the country. It was her mother's fault. The little girl cried, "No. You're not my mommy. No. You're not my mommy." He returned to his rocking. The mother whisked the child to the farthest corner.

Court officers jangled and thumped through the crowd with their arsenal of keys, handcuffs, and bouncing billy clubs. One wiry officer with a big smile, stationed outside Judge Gayle Roberts's courtroom, pierced the background buzz with a bellow worthy of Grand Central Station: "Step up! Step up! Anyone with a case in Part 1. Step up and sign in!" chanted Jack DeGiorgio with a wry twinkle. When the clock rolled on to 9:00 A.M., he switched gears to his Don Pardo imitation: "Party's o-O-ON in Part 1! The-party-is-ON!"

At once the atmosphere was transformed from dignified court-room into hectic train station. A cacophony of jittery chatter mixed with the smells of babies' apple juice filled the hall. It was a potentially lethal stew. Abusers sat next to their victims. Child welfare workers sat next to the parents they'd accused of neglect. Lawyers conducted interviews with clients they'd just met, discussing the most intimate subjects within earshot of at least two dozen strangers. One recent innovation was a child care center generously stocked with blocks, trains, dolls, coloring books, and snacks. Day care workers had to be extra vigilant to make sure only the legal guardians dropped off and picked up their charges.

Family court was a notion conceived by progressive Social Gospel disciples at the turn of the last century. These courts essentially replaced punishment with paternalism, removing juveniles for the first time in American judicial practice from the harshness of adult courts. As the century progressed, all manner of family disputes, from custody to neglect to parental rights, were decided in what were supposed to be intimate settings. The judge's bench was life-size and accessible. The legal language was altered to emphasize process over punishment. Courtrooms were called "hearing rooms." Trials were called "fact-finding hearings." Sentences were "dispositions," plaintiffs were "petitioners," defendants were "respondents."

The result, unfortunately, was often more confusion than coziness. The well-intentioned, family-friendly process quickly devolved into undignified chaos. New York City caseloads ballooned in the 1980s as the crack cocaine epidemic caused a parallel epidemic in child abuse cases. Thousands of poor children were placed in foster care. It fell on the family court to oversee both parent and child. Lawyers had won the right in the U.S. Supreme Court to represent children two decades earlier. Parents were offered less vigorous, but still gratis representation. What was protective of individuals' legal rights became bureaucratic bottlenecks for the courts. More time was required per case to gather all these adults to the conference table.

All this was happening at a time when New York City Family Court was starved for cash. Its buildings were crumbling, support staff was scarce. Judges were juggling as many as one thousand cases at a time, often sharing legal assistants with other colleagues. Up to fifty cases rotated through their courtrooms daily. Six minutes, on the average, was as much time as they could give to each one. No one kept strict records on how many of these cases involved people on welfare. But at least one 1998 study showed that families earning $15,000 a year or less were more likely to cross paths with the child protective system than those with incomes twice that amount. In fact, children like Christine's living in Morrisania were some of the most vulnerable in the city. One in eleven of the Bronx neighborhood's children were placed in foster care in 1998, compared to one in forty in the rest of the city. Morrisania's median income was $11,000 that same year. City-wide, the median income was $32,000.

By the mid-nineties, as the great exodus from welfare took hold, family court calendars became more jammed than ever. Between 1995 and 1998, the number of children removed from poor homes because of neglect or abuse increased by a full 55 percent. An unchanged number of family court judges (forty-three) were hearing 100,000 more cases than they had heard a decade earlier. Nationwide, the numbers of children removed from parents charged with abusing them dropped slightly after the welfare reform bill passed. The difference in New York City, however, could be traced to a change in the child welfare agency's priorities. When six-year-old Elisa Izquierdo's body was found by police in 1995, horribly brutalized by her own mother, the city overhauled its child protection agency to make saving children its top mission. Since that date, the number of children admitted to foster care had gone up from 8,000 to 12,500 in three years. The vast majority of the parents accused of neglect or abuse, like Christine, suffered from substance abuse. Only a tiny percentage of respondents, fewer than 2 percent, won their cases on the merits.

Christine was facing long odds.

An hour later, at about ten-thirty, Christine entered the crowded scene. She was late. Her hair was wet from a shower. She was dressed in a tasteful black dress and stockings. She was nervous about everything—about the fate of her children, about her own fate. She was anxious about facing Luis and enduring his disapproval. Life had ground down to slow motion for Christine after she lost her kids. She had grazed through the past week with a sense of empty dread. She was a sensual woman, one who thrived on hugs and touch. The physical absence of her kids made her crazy. When she tried to visit with friends outside, the clamor of their children playing would drive her back into her solitary room.

Dyanna spotted her mom and ran to her, leaping into her arms. Christine chattered excitedly with the foster mother, with Salimie Calim, the ACS caseworker. She kissed Dyanna on her forehead between sentences. Kristopher flipped up his hands to be picked up. Monica stood a little apart, looking wounded, and likely scared. "You're late," Luis pointed out. "What else is new?" Stung, Christine tried to ignore his rebuke.

Just then, a soft-spoken man wandered by rote through the benches, calling out "Christine Rivera." Philip Francis, an elderly gentleman with gray-flecked hair and four grown children of his own, was one of the city's three hundred private attorneys hired on the cheap to represent indigent parents. In many ways the parents got what the city paid very little for. These "18-B" attorneys, named after the section of law that created their line of work, earned only forty dollars an hour for court work and twenty-five dollars an hour for out-of-court time, fees that had remained unchanged by state law since 1986. Like most 18-B attorneys, Philip had more than one hundred parents in his caseload at all times—too many, he acknowledged, to do them all justice. On average, his fellow attorneys won their neglect hearings less than 2 percent of the time. One reason was that judges had only to decide on the "preponderance of the evidence," not beyond a reasonable doubt. Another reason was simply that 18-B attorneys did trial preparation on the fly. With no staff, no team of social workers to help their clients, they had little time to track down witnesses and build the strongest case. Many spoke to their clients for the first time in a whisper before the judge. Others carved out five or ten minutes in the waiting area. All this led to an inevitable sense of futility. "I tell my clients, yes, we can go to trial," said Philip Francis. "But we will probably lose. That's just how it goes."

Federal law had made the court's job even more onerous. A new adoption act imposed a deadline on the courts to decide a family's fate. The intent was to make certain children did not languish in the impermanence of foster care for years while the bureaucracy lumbered and delayed. Judges now moved to terminate parents' rights if the child had spent fifteen of the past twenty-two months in a foster home. The fate of parents like Christine, who might require a cure that extended beyond the deadline, was unclear. "This is a sad, sad court," said Philip, glancing at the foot of casework on his lap. "It's even sadder now than it has been in years."

Christine gently dislodged Dyanna from her leg before meeting with Philip. Dyanna scuffled toward her seat, dejected, throwing her head down on the back of the bench in convulsive sobs. In another room, Christine told Philip that she knew about all the charges in the ACS complaint. Yes, she did skip her rehab program on occasion. Yes, her urines were dirty a couple of times, but not always. Yes, she sometimes left Monica and Dyanna alone in the shelter, but only briefly. She always knew where they were. She gave him a letter from the shelter doctor explaining that some of her absences were due to Kristopher's illnesses. Still, Philip decided not to contest any of the city's points. He had neither the evidence nor the time to build a case. Besides, Christine's drug lapses doomed her chances of getting her children returned in the near future. The best he could do amounted to damage control. He would try to gain visitation rights for Christine, to make sure the court followed the law. Their conversation took eight minutes.

When Christine emerged, the foster mother was packing up the baby's things to leave. No one was waiting for the judge's decision. It all seemed preordained. Christine bent down to kiss Dyanna one last time, whispering to her, "I love you. You'll be safe there. I'll see you soon." Dyanna's sobs grew frantic. She wrapped her legs and arms around her mom, as if to anchor her in place. Monica inched forward, blinking tears from her eyes. She gave her mom a reluctant hug, and then hot tears streaked her face. Christine reached to pick up Kristopher, who wanted in on all the commotion. All three children surrounded her, clinging, frozen in a moment.

Two adults gently worked to unravel Dyanna's limbs from her mother. Kristopher wailed as someone unglued his grasp. The foster mother moved toward the elevator with the three sobbing kids, Kristopher in his stroller.

Then Christine's children disappeared from sight.

A spontaneous quiet rushed across the benches. The scene was unusually bitter, even for family court. People embroiled in their own private catastrophes stopped to grieve for a moment at the sight of this unknown mother losing her children, and her children losing her. It was everyone's worst fear—to have your own kids pried from your arms, wailing, sobbing, then handed over to a stranger.

Christine shuddered, trying to suppress her sobs. The crowd's attention made her uneasy. No one was there to comfort her. Talk helped her stop the tears. So she talked to her neighbor of the moment. Would the girls be safe? When would she be able to see them? She knew that losing her kids was largely her own fault, but she never anticipated this. She brought out a scrap of paper with handwriting on it from her purse. It was from Graham Wyndham, the city-contracted private foster care agency now in charge of her kids. There was a contact name and phone number, and her kids' names and ages. It looked like a handwritten receipt for her children.

Luis stiffened at the room's opposite end. He was a knot of anger, staring at the ceiling. In all his years working as an interpreter in criminal court, he had never witnessed such heartbreak as this. "This is a zoo," he said, and walked away.

By the time she was called into Judge Roberts's chambers nearly two hours later, Christine was numb. The hearing seemed little more than a poignant afterthought. Everything appeared to be moving along someone else's well-oiled rails. Her attorney was treating her case as more or less routine. Christine stood before the judge, her hands behind her back. She gave her name and pledged to tell the truth. That was all. After that, she felt herself disappearing. No one asked her questions. Judge Roberts barely acknowledged her. There was talk of visitation rights. Luis would get Kristopher. The girls would register for another school. One detail faded into the next. The next court date was September 23, 1999, an entire summer away. The only thing anchoring her to a life worth living was slipping away. She felt invisible.

13

Alina

Porridge and Champagne

Women like Christine, living on life's edges in the Bronx, faced setbacks and aggravations familiar to those living in far poorer nations, in the spoils of the former Soviet Union. The difference was that New York women shared the sidewalk every day with the middle-class, the wealthy, and the exceedingly wealthy. Affluence was the norm. Poverty was a shadowy, societal by-product. The experience pushed them further into a life of otherness, of less-than-ness. Their American world was, in this way, more undemocratic than that of the former Communist nation of Moldova, Alina's native land, where virtually all the inhabitants shared a similar, impoverished fate. If Alina had forgotten this, then she was reminded quickly enough when she returned, with considerable trepidation, for her June visit. The minute she arrived in Kishinev, Alina became both a witness and sympathetic ear to the tales and travails of old acquaintances.

Visitors lined up each night to see the Americanized Alina as she held court in the apartment of her friend, Natasha Isakova. These were old friends and former neighbors of Alina's, of her mother's, her grandmother's, of Mark's relatives. Alina greeted her guests with gifts of five or ten U.S. dollars—much more stable currency than the Moldovan *lei*—plus a hand-me-down sweater from a New York thrift shop. In turn, the guests brought what they could: homegrown berries, jam, store-bought cakes. Others who had nothing to spare brought only apologies, requests, and stories of living hand to mouth in a bribe-driven economy.

These Kishinevites had not fled Moldova, for reasons that said as much about their genealogy as their nationalist pride. The final irony for Alina was that all but one—an ultra-nationalist Moldovan—now wanted nothing more than to uncover just a trace of Jewish blood in their Russian or Moldovan lineage. That might be enough to justify a refugee visa. Jewish philanthropies and the Israeli and North American governments helped Jews, but very few others, flee the economic disaster that was Moldova. For the first time in Kishinev's five-hundred-year history, being Jewish was a desired state, tantamount to salvation.

On one evening typical of Alina's celebrated stay, Natasha Isakova returned from her nursing job, wondering how much respite she would have before the doorbell began ringing. She picked up the phone to make some arrangements. An airy *whoosh* had replaced the dial tone. The phone's low growl hadn't rung for hours.

She was already living without hot water, without nightly electricity, and now, without a phone. Electricity and phone service—these were the amenities the poor in America also lived in dread of losing. Yet Natasha was a professional. She had a full-time job, and her grandmother's inheritance. She dashed back out into the rainy night to find some answers. Days earlier Natasha had given her sometime boyfriend Sergei money to pay her delinquent bill. He was manning her minivan that night. The only way to reach him was to stand at various unmarked bus stops and flag him down. She didn't suspect Sergei of stealing. Maybe he forgot. Maybe the phone technicians just didn't do their job. Or maybe she just wanted an excuse to see her boyfriend on a cheap date and to escape the bustling reception center her American guest had created.

The first visitor was Alina's second-best friend, Natasha Gargan, who was living the life as a native Moldovan that Alina once envisioned for herself as a Moldovan Jew. The honey blonde in jeans and white tank top was starting her fourth year of medical school. In one year's time she would be starting her doctoral residency. Gargan was about to set up housekeeping in her parents' apartment with her fiancé, a friendly young shoemaker dressed in a made-in-Romania Nike T-shirt. Few newlyweds could afford to live on their own. Natasha Gargan unwrapped the jelly cake she had brought to fete Alina's homecoming and pulled out a photo album. One picture showed Gargan as a young doctor-in-training, dressed in what looked like a tall white chef's hat. It was regulation maternity ward uniform. Gargan was cradling a newborn in her first internship at the baby hospital, the same one in which Alina's brother Moshe had been born.

Gargan explained that the Kishinev Medical Institute might be free on paper, but it was subject to the same system of bribes that energized and polluted every other facet of Kishinev's economy. Each student, for instance, was expected to come up with a five-hundred-dollar bribe to enter the school. Another two thousand dollars slipped to the right person would buy a decent position as a physician in residence—a fortune for the average Moldovan. On top of that, each professor expected a bottle of Moldova's export-quality champagne, or some other special gift, before grades were distributed. It was often the instructor's only steady source of income. The bartering became frenzied as the stakes grew ever more intense. And the baby in her arms in the photo? Natasha said she knew of people who sold similar babies, those born unwanted, to the highest international bidder. The going price was three hundred dollars.

Even after hearing such stories of corruption and uncertainty, Alina still felt a twinge of regret. If she had stayed in Kishinev, she would be halfway to her goal, alongside her friend. As it was, she had no idea whether she would ever be accepted into an American medical school. Despite her strong will and stellar undergraduate grades, she did not have the $24,000 annual tuition, and her MCAT scores were on the low side. On top of that, she trembled with fear in all her English-language admission interviews. No school, she thought, would risk accepting such a tin of nerves as a future physician.

Alina's struggles were hard for her Moldovan friends to comprehend. She had much more money per month than any of them, even though her welfare check kept her well below poverty level in the United States. She had a future in the world's most prosperous nation. But her struggle to reach her goal was compounded by the gaps in her income and language. Her friends peppered her with questions about the size of American kitchens, cars, streets, buildings, NBA basketball stars. Money and size were as relative as poverty. Her friends could not fathom the huge cost of living in New York, the out-of-sight tuition fees, or the hours wasted working WEP when she should have been studying. They couldn't understand the stress of living as a welfare recipient in the richest country in the world, one that would never feel like home. There was no use trying to explain.

Gargan bolted from the kitchen bench, remembering her medical school exams the next morning. She hugged Alina, promising to return before her departure. A second visitor was already waiting in the entryway with her young daughter. It took Alina some time to realize that the red-faced Moldovan woman, with a bright smile and leathered hands, was once the Zukinas' neighbor in the coveted Soviet apartment complex. She looked worn out from the last four years of scraping to find work and food. Her husband was a shoemaker who couldn't afford new shoes. Her situation was so stark that even her dog was starving. The dog used to live off the scraps left by the kindergarten children in the school where she worked on and off as a cleaning lady. Now the school could not afford to provide either meat or porridge for the children's meals, only tea and dry bread. The children lapped up every crumb.

Alina was embarrassed at the woman's effusive gratitude for her gift of five dollars and a used shirt. The woman apologized profusely for having nothing to offer in return. "I don't have tea or eggs, just beans. Can I give your mother some beans?" she asked Alina. "My mother has enough beans, thank you," Alina replied, feeling worn out herself from the tide of hopeless news.

While Alina was talking to her mother's old friend, another visitor paced in the hallway. The attractive young woman was dressed in a long rayon skirt and blouse, wringing her hands, anxiously awaiting her turn to hold audience with "the American." Alina had never met this woman before. She was a friend of a friend, so desperate to escape Moldova that she carried her appeal to a stranger. Her voice cascaded up and down in anxious tears as she explained her situation, fumbling through her bag for a professional photo she brought of herself. In it, she was dressed in a white frilly blouse and black skirt, posed at an awkward angle on a large boulder, her bare legs angled provocatively, her smile voluminous and red-lipped. She explained to Alina that she would like her to take this photo back to America and post it on a web site that specialized in matching European men to foreign brides. She figured the only way she could escape Moldova was to marry outside its borders. Anything was better than sewing at home for twelve hours a day, making barely enough to buy porridge for her brother and her mother. She had run out of money for the bribes her brother needed to sustain his grades in college. In this backwater country, the young beauty had no hope of ever meeting a decent husband who could offer her another, more lucrative citizenship. She had nothing worthwhile to sell but her marriageable status. A Moldovan "Picture Bride."

It shouldn't have surprised Alina that this young Moldovan woman would essentially put her single life up for sale on the Internet to the highest bidder. All of Kishinev appeared to be in a selling frenzy, perhaps to make up for fifty years of state-ordered vending bans. Moldova produced very little for export, besides fruity wines and chocolates. Its agricultural and military industries had collapsed with the Soviets' departure. Even the capital's Bucuria chocolate factory was forced to shut down in the summer because the chocolate would melt in the heat. It couldn't afford air conditioning. So the citizens turned into ad hoc vendors to earn a few *lei*. In New York, unlicensed vendors were part of the underground economy, fined and arrested if caught. In Kishinev, they were the most public and prolific commerce, making tiny deals to provide basic sustenance.

The sidewalks of the grand Boulevard Stefan cel Mare became an eclectic flea market during the day. Rickety tables were filled with odd assortments of Romanian cigarettes, sunflower seeds, and toilet paper. Lone vendors set up little bathroom scales or large, rusty, hospital-sized ones in the middle of the boulevard, charging one-quarter of a *lei* (two cents) for the privilege of being weighed in public. In a different business venture, a lineup of housewives stood poised, shoulder to shoulder, holding knitted trinkets, socks, and bras draped over their arms for sale. This is what Alina's mother had done over the border in Romania, before the family emigrated to America. The underground market in Kishinev offered other kinds of goods and services, such as phony Canadian immigration documents for three thousand dollars. American ones sold for five thousand dollars. Natasha Isakova joked that the sales fever was so bad that even the Kishinev opera house peddled furniture during intermission. Since few people had cash to spare, it was an economic stalemate that left little hope for progress.

The outdoor Kishinev food market provided the city's biggest industry. Blocks of farmers' fruit and potato stands lined the lot, with Ricky Martin's "Livin' La Vida Loca" blasting over the loudspeakers. The indoor House of Cheese sold table after table of the white, home-made *brinza*, a soft, dense, ricotta-style cheese, heavy on the salt. Next door was the profitable House of Meat, and then the House of Lard— or *sala*, the tasty pork slices used for cooking grease that Alina insisted could not be found in New York. Alina noticed first the shoppers' flamboyant attire—women in gauzy evening dresses and high heels, men in suits. "They all look as if it's their last day on earth," she said, wondering from where this sartorial display derived.

Alina shook herself from these market thoughts and accepted the woman's provocative photograph. She promised to try and find the right web site for her. The woman sobbed a little, and left. These visits left Alina shell-shocked in many ways. She did not expect to find that her family's small monetary gifts would translate into two months' salary. She was filled with a sense of sad finality. The country she had dreamed every day of returning to was the place all these friends still wanted to flee. She felt more drawn to its smells and tastes than those in New York. She preferred its leisurely pace. She was more comfort-able in a city where people could wear the same clothes several days in a row and no one would care. Where she wasn't just another WEP worker among thousands. Kishinev would always be home. But clearly, it was disintegrating.

Just then, Natasha Isakova returned in a wet flurry. She was soaked to her skin. The black, knock-off Adidas sneakers she'd just bought in the market that day had bled through to her white socks, turning them gray. Natasha had spent her Friday night riding around Kishinev with Sergei in the minivan, picking up passengers. Sergei would pay her phone bill tomorrow. He had forgotten. The phone would be back in service by the end of the week. She'd also learned on her nightly foray that her father and stepmother (a woman with some Jewish blood in her heritage) would soon be emigrating to Israel. Natasha would be left behind. She could not claim a bloodline. Her late mother was Russian.

The visits to Alina were not quite over. At 7:00 A.M. the next morning, the bell rang insistently. In many ways this last visit was the most extraordinary. Alina stumbled to the door to find her mother's old colleague, Nicolae, standing stiff and proud in the stairwell, his bronzed face obscured by the enormous bouquet he carried in plastic wrap. He smothered Alina in a bear hug, tickling her with his brushy mustache. She struggled to compose herself as he hauled in bag after bag of gifts: pounds of the finest *kielbasa,* a sausage she hadn't tasted in four years; huge slabs of *sala,* the lard the Zukinas craved; cheese, and more sausages; followed by four hundred *lei,* more than she could ever hope to spend in the few days she had left. "For your mother," Nicolae said, standing in Natasha's kitchen with his smartly coifed gray hair and authentic Adidas red, white, and blue sports outfit. When Alina protested over the excessive gifts, he said, "No problem. My wife is the director of the House of Meat." Alina knew this to be a very important position in the market.

"Come, I have something to show you," he said hastily. Alina quickly stashed the used sweater and ten-dollar bill she had laid out when she heard Nicolae was coming. She hoped he hadn't seen it. She followed him out to his leather-interior Audi parked in front and climbed in for a drive through the streets of Kishinev at what felt like sixty miles per hour, swerving around potholes, gunning for pedestrians. When a traffic officer in a cowboy-style hat waved him down from the side of the road, Nicolae dashed out of his car, pulling cash from his pocket. He returned one minute later laughing, "I told him who I was, and he waved me on."

Exactly what he was, was still a bit of a mystery. Nicolae had worked in the cafeteria with Alina's mother, Lyusya, many years ago, cooking, serving, just like everyone else. As a side job, he chauffeured official people around, making acquaintances in high government places with his gregarious personality and fearless diplomatic skills. Perhaps those connections had translated into newfound riches when the Soviet Union collapsed. Anyone with this much cash was usually suspected of illegal activities. Yet it was still guesswork on Alina's part. Nicolae was not as quick to identify the source of his wealth as he was to shower it on his old friend's daughter from America. "I'm good to the good guys and good to the bad guys" was as much as he offered by way of explanation.

He continued on a country road that rolled lazily up and down the green farmland, reminding Alina that Moldova had always been primarily an agricultural country. The Mediterranean-style climate produced meaty and aromatic grapes, tomatoes, and strawberries that could perhaps be someday organized for export. Early cornstalks sprouted six inches high along the road. Small plots of land dotted with lean-to shacks marked the spots where Kishinevites cultivated their own personal gardens—often providing families with their only source of berries and vegetables for the year. About ten kilometers down the road, Nicolae pulled into a little village called Valei Kolonitza, a former penal colony, and turned right down a road so craggy even he had to slow down. Herds of goats, cows, and geese occasionally meandered across the gravel.

"We're almost there, almost there," Nicolae said impatiently. Rounding the corner was his destination. Here was his own private monument to all that was uniquely, and organically, Moldovan: a three-story limestone house with red tile roof and teak woodwork, representing his personal protest against half a century of aesthetically ugly Soviet rule. The home's terraces and angled windows perched boldly along the side of a small hill. A freshwater well provided water. An organic garden was budding with early summer vegetables and flowers behind the house. A generator guaranteed that the finished mansion would be immune to the nationwide utility meltdowns. "All Romanian materials," Nicolae boasted. "All natural. All organic. Not that plastic Russian stuff," he scoffed.

He volunteered line-item costs. The woodwork cost 75,000 *lei*. The wooden outside staircase, 70,000. The trim alone came to $15,000, a staggering amount for the average Moldovan. It was a breathtaking display of ostentation amid deprivation—one that left him beaming with patriotic pride.

The unanswered question still hung over the morning. How could Nicolae afford all this when most of his countrymen were scraping by for mere survival? His story began to trickle out in spurts as he poured out two bottles of fruity champagne and one liter of home-brewed red wine ("a gift from the minister of agriculture") one glass at a time. "I only drink for my health," he repeated often. Alina struggled to keep up on the imbibing front, but it was early, and the midmorning sun was attracting steam and flies to the enclosed garage. "I don't drink vodka or cognac, not like the Russians," Nicolae said several times. He said his past KGB work during the Soviet years secured him a plum job after the revolution. He said his job was to chauffeur Kishinev's food market director, probably the most financially powerful man in the capital. "It's a dangerous job, driving for such a wealthy business-man," he said, as he described a world that sounded much like the lawless Wild West, Romanian style. Racketeering was rampant in the market. Vendors had to pay twice for their stalls—once to the director, once to the racketeers, who were protected by the police. Someone had attempted unsuccessfully to assassinate his boss by wiring his car with a bomb. Every day, thieves tried to run off with the market cash.

Nicolae stopped short of explaining how these connections resulted in his unfinished real estate masterpiece. He believed he deserved such a house, as a patriotic Moldovan. He regretted that the nationalist fervor that gripped his tiny nation in the early nineties had since abated. He blamed its dissolution partly on the Russian wives of the past two presidents. As long as Russians maintained a toehold in Moldova, he said, it would never progress. What was happening now was simple justice. "Russians stole Moldova from the Moldovans," he said. "They imposed their language, their politics. They took the best jobs. Now it's the Moldovans' turn." As for the Jews? Nicolae said he considered Jews more Moldovan than Russian. Jews were historically farmers in this rural land, not urban bankers or businessmen. Thus, they had blood ties to the earth, just like his ancestors.

It was not clear whether Nicolae was saying this in deference to his friendship with Alina's family or whether he believed it to be true. What was clear was that fresh produce was central to the Moldovan experience, as was an odd farming fervor. Nicolae's xenophobic rhetoric explained why the independence movement had taken such a firm hold in the first place. And his nouveau wealth explained why Alina probably would never in her lifetime see her beloved Moldova climbing out of poverty to an economic middle ground. With piles of *lei* in the hands of the privileged and connected, the situation seemed hopeless.

Alina flew off to New York the next day, feeling a new gulf between her life and the one she was leaving behind for the second time. The visit had filled her with conflicting jolts of nostalgia. She missed everything about Moldova, most of all the forgiving sense of closeness and well-being. This was home, no matter how uncomfortable, no matter how volatile its history. But she knew that she could never return. It was hard in America, but now she was more resigned to it. In many ways she was more prepared for the struggle off welfare than her American counterparts. She had tasted a form of poverty in Moldova that was not splashed with the stigma of race, class, or personal failure. She walked into New York's welfare offices with a strong education, a strong family network, and a sense of pride in her past accomplishments. She would trade this sense of belonging for a future with some hope.

Besides, there was the mail to look forward to in the Bronx. She might have a letter from HRA cutting her off welfare for the third, and probably final, time. It was inevitable. But what about her next life? Would she find any acceptance letters to medical school?

14

Brenda

Wall Street

A man in a giant chipmunk suit chuckled and pumped the hands of tourists at the gateway to Wall Street. It was a few days before Christmas 1998, and even the world's most formidable financial district twinkled with festive anticipation. Brenda gave the costumed creature a pleasant nod as she strutted past. Her hair was stylish Supremes, puffed and wavy. She had scrimped for months to save seventy dollars to pay for the elaborate hairdo. The going rate was one hundred dollars, but the Bronx beautician agreed to give Brenda a deal. Two women required all day to weave such a creative hair sculpture. It was worth the trouble and expense for Brenda. She felt like looking good today. This was her first day as a permanent Aramark employee, no longer on probation as an America Works trainee. That day before New Year's marked the end of those numbing welfare lines and humiliating home visits. "Never again. Uh-uh. Not those lines," she hummed to herself.

Something about her head-high swagger made the security guard at 60 Wall grin in spite of himself. Exactly one year ago she was wondering how she would tell Ty that Santa's sleigh must have broken down before he got to the Bronx. This Christmas she could decorate her tree with the toys of her labor. Everything was falling into place. Loreal was a freshman at Brooklyn College. Ty was finally enrolled in the Head Start preschool. An Urban Horizons neighbor watched him until Brenda could get home. Arriving at work, she went down the long escalator, through the maroon-and-white-tiled cafeteria, past the fruit baskets, the gourmet coffee station, and headed for the glassed-in office. There it was. Her own time card with her name and photo printed on it. She flipped it through the computerized punching machine and laughed out loud.

"I wish I had a thousand Brendas," said Robert Gordon, the cafeteria manager, when he handed over her first paycheck with a flourish. She doubled over in a giggle and headed for the employee locker room. Robert had seen a spark in Brenda four months before, when he took her on as an America Works trainee. "She had heart," he said. "I'd rather take someone with heart over someone with work knowledge any day. I can train anybody. But I can't teach you to smile and to have desire." Brenda had a way with customers. Customers got a "Hello, how you doin'?" and a "Let me take that back for you" when their juice spilled all over their sandwich, or when a fish steak wasn't quite cooked to suit them. She made time to ask about their kids, their health, their job worries. Before long the lines at Brenda's cash register were twice as long as those at other registers. Cards and gifts poured in for her on Thanksgiving, Christmas, and later Valentine's Day. Even the vice president of J. P. Morgan offered his help and his friendship, if she ever needed it. The attention often rankled the other women who worked beside her. Brenda dismissed them as trifling.

After a couple of months Robert asked Brenda whether she would help him out by working the night shift, locking up the cafeteria at least one night a week. The 1:30 to 10:00 P.M. hours would be hard on

Ty, and on her sleep, but the request made her feel good. Robert valued her work. Otherwise, he wouldn't ask her to take on the extra responsibility. She weighed the request for a day or two. Brenda had witnessed some testy scenes during her probationary period. Several higher-paid coworkers cried when they got their hours cut in half, a tactic designed to move them along. Brenda, earning just $5.15 an hour, was being asked to work overtime. One male coworker tried her patience with his sexist needling. When she sponged down the salad bar, he'd sidle up behind her and say, "I love a woman who can scrub." When she passed the grill station, he'd make some "fresh meat" remarks. Finally, Brenda hauled off and lectured him. "I'm not losin' my job for you or nobody," she told him. "I love this job. I need this job. I'm gonna get me my respect. Back off."

The aggressive tactic worked. It could have backfired, but it worked. The taunter retreated. This sealed her don't-mess-with-me reputation. Meanwhile, her loyalty to Robert paid off. She worked the night shift for him. He gave her the manager code to close up the new $6 million cafeteria by herself. He hired her at $8.25 an hour, more than most with her level of experience.

Robert Gordon was a transition man, trained in the old family-style way of business at J. P. Morgan, now learning the new corporate approach under Aramark. Under Morgan management, the food had been free, while the cafeteria employees earned around fifteen dollars an hour. Under Aramark, customers paid discount prices for their meals and employees' hourly wages were practically cut in half. Robert was overseeing the often wrenching changeover. About fifteen people were either laid off or transferred under his watch. The staff dwindled from one hundred to seventy. Robert saw a future for Brenda, however. "I see her as a supervisor one day," he told me. "She's irreplaceable."

Brenda's starting salary placed her well above the average New York City welfare mother returning to work. HRA reported that more than 40 percent were earning only $7 an hour or less. Still, it wasn't long before Brenda realized that even $8.25 an hour was not going to buy her vacations in Disneyland, let alone textbooks for Loreal or shoes for the sprouting Tyjahwon. She took pride in her work. She could see herself moving up to management in a few years' time. But with this wage, she was eking by.

An economic study was released around this time saying that a working parent in the Bronx with two children needed to make $38,000 to live self-sufficiently. That sounded about right to Brenda. A nagging sense of static began pervading her life. Her gross salary came to about $16,000 a year—a full $1,500 more than the federal poverty level for a family of three. Her monthly cash flow amounted to about $1,100 after taxes. Out of that, $400 went toward rent. (Federal Section 8 was still picking up the other $65.) Child care cost her $240 a month because of all the late hours. The subway costs to transport herself to work and Loreal to college came to around $120. The phone cost her $50 per month, utilities $35. Groceries and medicine came to about $300.

Already the bills just to survive added up to about fifty dollars more than she brought home. And she wasn't counting clothes or school supplies. She was still cashing her checks at the ubiquitous check-cashing storefronts, just as she did when she was on welfare. There was never any money left over at month's end to open a bank account. She paid her bills by hand every month, walking cash over to the telephone company, to the electric company, to the rent management office. "I'm always behind," Brenda said. "Always strugglin', strugglin'."

The welfare reform law allowed for six months of transitional help as welfare recipients moved into low-paid work. Lawmakers realized that large numbers of the formerly dependent would probably move into jobs that did not provide health insurance or enough salary to fund child care costs. The extra help was meant to stave off a national health crisis for both kids and their parents. Not everyone knew these funds were available. Brenda was an exception. She kept her Medicaid insurance for the first few months. Many others went without altogether. In New York, 265,000 people dropped off the Medicaid rolls after the welfare reform law took effect, representing a 9 percent dip in 1997. Those receiving food stamps dropped off by the same percentage. Nationwide, one Families USA study estimated that 675,000 lost health insurance completely in 1997 when they moved into low-paying jobs. Most of these people were children.

Child care was another, more tangled story. Brenda thought she was eligible for about $2,000 for the first year of work to help defray her baby-sitter costs. But after months of applying, she got word that the city would refund only $35 a month—$200 less than she shelled out. That was hardly worth the hours she spent filling out the forms. The federal food stamps office told her she made too much money to receive help buying groceries. "They used my gross, which doesn't seem right," Brenda said. "They also figured I was getting child support, which rarely came steady." Ty's father was required to pay $374 a month in child support directly to welfare, which he did irregularly. Then the city would give Brenda $50, on top of her welfare benefits, and keep the balance of Teddy Jenkins's payments to help subsidize Brenda's assistance. Once she closed her case, then she would directly receive the entire $374 from Teddy.

After several months Brenda's Medicaid coverage timed out. Loreal found that out when she went to the doctor for an eye infection and couldn't pay the bill. Brenda was under the wrong impression that they'd be covered for a year. Ty was insured under New York State's impressive Child Health Plus for children under nineteen. But the two women went without coverage for months. Aramark offered health insurance, but it required employees to kick in sixty dollars a month to cover the benefit. Brenda couldn't afford that. Her blood pressure medicine lapsed just as her stress levels mounted. She had trouble sleeping. Her heart would race. Then, just when she could least cope with it, baby-sitter problems hit hard, once again. Ty told his sister that a young boy in the baby-sitter's house was touching him the way he didn't like. Brenda confronted the mother at midnight the same night and vowed never to let Ty enter that apartment again. It took her several more weeks to find someone else she could trust.

At the end of the year Brenda thought a possible $3,500 tax refund might give her that extra nudge into solvency. She was counting on it to pay her March rent and to catch up on other bills. It was money from the vaunted Earned Income Tax Credit, a federal program designed to reward work by offering large returns to low-income workers. Conceived as a "work bonus" under President Richard Nixon in the mid-seventies, the tax credit for the poor was hailed by Ronald Reagan a decade later as "the best anti-poverty, the best pro-family, the best job creation measure to come out of Congress." The tax measure lost a little of its Republican support in the late nineties when Clinton succeeded in winning a considerable boost in the return rate. But as long as EITC was never stigmatized as a "welfare" program, it had great potential to rise above ideological stalemates and to move the poor into life without debt. In 1996 alone the working poor tax credit lifted 4.6 million people above the poverty line.

As it turned out, Brenda received the maximum tax return, but she never saw any of it. Her windfall went straight into paying off a state loan she had taken out more than a decade earlier for tuition at a trade school. Mandl Business School had promised a job as a medical or dental technician for life, but it didn't work out for Brenda. She walked out of an internship with a North Bronx dentist feeling demoralized. For one thing, she was left-handed, and the tools were made for righties. For another, she was African American. "One patient wouldn't let me touch him because I was black," Brenda said. "That's when I left and went into food service." That loan had hung over her future all this time.

Meanwhile, Loreal quit Brooklyn College temporarily and started working at the Wiz electronic store—partly because she was tired of being a drain on her mother, partly because she was tired of keeping up with her studies. "My family is getting away from me," Brenda said one weekend in her apartment. "They are using me up at work. I'm the only cashier in there at night." The pungent sting of bleach filled her kitchen. She was soaking her white uniforms in the sink. Aramark's laundry didn't whiten them enough to suit her standards. Stacks of folded laundry filled every corner of the couch. "Ty is suffering, always sick. I have to wake him up at the baby-sitter's at eleven o'clock every night to bring him home. He don't have a life anymore."

Brenda had entered the ranks of the working poor, a labor statistic on the march in America. Two-thirds of the nation's welfare families had left the dole for jobs by 1997, but one-fifth of those were worse off financially than before. By the millennium, a full one-third of New York City's families fell below the poverty level—twice the national average—despite a rise in employment and education. It was becoming depressingly clear that a job was no guarantee that a family would escape poverty.

Three million more American children were poor in 1997 than two decades earlier. For every five children in America, one was living in poverty. It was a humbling record for the richest nation in the world during a period of unstoppable economic growth. No industrialized country could claim worse. The latest trend in 1999 was even more sobering. Two-thirds of those 5.2 million poor children lived in homes where one parent was working, according to Columbia University's National Center for Children in Poverty. That represented a 42 percent gain in the numbers of working poor families in twenty years. Seventeen percent of those children were black, 24 percent were Hispanic.

Hardships took their toll on life's staples. In 1999 one of the first comprehensive surveys to codify the millions dropped from the welfare rolls nationwide found that one-third were skipping meals for lack of food. Nearly half of the respondents told researchers they had trouble paying their rent and utilities. One-third said they were worse off than when they had been on welfare, according to these findings from the Urban Institute, a nonprofit research group. One-quarter were not working at all—the truly missing people. Nearly 30 percent returned to the rolls within the year.

Reality was more difficult to quantify with any accuracy for New York City. In the six years since WEP had been launched, HRA had released only one minor study tracking what happened to the half-million people slashed from its rolls. Critics immediately pounced upon its limited scope. The report, called "Leaving Welfare," surveyed only 126 out of a total of 6,000 cases that were closed in one month. In addition, its pool was limited to those who had telephones, which skewed the results to the most successful former welfare cases. Still, the city study found that only 40 percent left welfare for a full-time job. About 15 percent left for low-paying part-time jobs. Another one-quarter were kicked off the rolls for violating its rules. Forty percent of those working were earning less than $10,000 a year, far below the poverty rate. One-quarter of those working, like Brenda, were working the night shift, putting extra burdens on their young children. One-fifth had less money than they did on welfare.

When Brenda said, "My family is getting away from me," she echoed every working parent's lament across all socioeconomic divides. It was an intangible refrain that floated uneasily beyond the scope of social science's measuring sticks. Middle-class mothers complained about missing their children's first steps, or their starring roles in the school play. Wealthy families worried that their teenagers weren't getting the parental attention they needed. The same concerns only deepened when hand-to-mouth incomes tugged at the family dynamic. If Brenda missed work for Ty's colds, her paycheck was docked. It was impossible to take time off, even for the short term, to shore up a child who was floundering in school, or a young teenager hanging with the wrong kids. She was just one paycheck away from the social services line each month. Her education level made it unlikely that her paycheck would ever improve. The main thrust of welfare reform, "work first," didn't allow for the stress of juggling home with work, let alone advancing one's education. "Work first" meant exactly that. The hope was that work would improve the overall quality of family life. It probably would, if the pay were better.

Every day WHEDCO's Head Start director witnessed the collision between the domestic demands of poor parents and the requirement to work. Head Start, a preschool program that served below-poverty families, was finding it difficult to sustain its traditional curriculum. Head Start's successful thirty-year record was built on the practicing theory that parents were active volunteers in their children's school. Rose Rivera lamented that such an objective was impossible to maintain after welfare reform. The parents were now working either WEP or minimum-wage jobs. Some were working WEP within the walls of WHEDCO, doing maintenance, kitchen, or handyman-type labor, which changed the traditional relationship between parents and their child's Head Start school. Rose was forced to rearrange the fundamental mission of her program. "The role of parent is undervalued now," Rose said. She could no longer require parents to participate. "They are tired from working day and night. Their children are tired from being institutionalized all day." While his mom was out earning a minimum wage, Ty took sound naps every day when he was in preschool. His days were as exhausting as Brenda's.

By spring, Brenda decided the night-shift work had to end. About the same time Robert Gordon announced he was being transferred to another cafeteria. Both moves would leave her chin to shoulder with a clique of daytime coworkers she didn't much care for, and a new boss who was not as sympathetic as Robert. Still, she figured it was important for her to be home for Ty when he returned from kindergarten. He would be attending the brand-new public school on the corner, the Rafael Hernandez Dual Language Magnet School, which formed the final link in the Morrisania Hospital renaissance. She had fought to get him in there. He would be one of the few English-only black children in the nine-hundred-seat bilingual elementary and junior high school. Brenda needed to make sure he did his homework and that he went to bed on time with a bath and a book. "He's doing good in school, but his manners are getting wild," she said. Brenda also needed to nudge Loreal back to Brooklyn College in the fall. "I always regret not getting higher education. She knows it's important."

The day shift turned out to be as testy as Brenda imagined. She was still doing the same kind of work—making gourmet coffees, stocking the salad bar, running the cash register. But the job that had once filled her with a sense of purpose now filled her with stress. Management was quicker to scold instead of promote. Coworkers were more likely to mutter an insult than say hello. As the millennial New Year approached, Brenda would head off to work every morning with a knot in her stomach and a prayer in her head: "Please, God, don't let me explode on no one today." It was getting harder to keep her cool, and her dignity. "They are treating us like dirt in there. It's a bad atmosphere."

One day it was raisins. A coworker accused her of leaving smashed raisins in the linoleum grooves at her workstation. Brenda popped off at her, saying the woman knew full well who left those squished raisins on the floor, that it was somebody else entirely. Manager Owen Moore called her into his office to lecture her about cursing out her coworkers. Another day she was accused of arriving at work a few minutes late. That time, she said, she went off on another manager, telling him he "talks to people like they are nothing." She ended up getting suspended without pay for three days. The suspension was later reversed. "I worked too hard to lose this job like this," she said. "This job is taking me out of my character. I'm having a harder time getting along in there than when I was dancing in bars."

In the midst of the unrest, union organizers began moving into the 60 Wall food court. The AFL-CIO Local 100 Hotel and Restaurant Employees Union had fought and won three other contracts with Aramark cafeterias in the city—at Salomon Smith Barney, Chase Manhattan Bank, and Metropolitan Life Insurance. Those battles were long and bitter. For a year, Aramark management at Smith Barney refused to recognize that the majority of its cafeteria workers had signed cards expressing their desire to have union representation. Local 100 President Henry Tamarin accused the employer of launching "a campaign of intimidation and harassment that made it impossible for a fair election." Aramark dismissed a cashier who was a strong union advocate and transferred two others to other cafeterias. Fearing a National Labor Relations Board dispute, Aramark eventually paid back wages to the fired employee and gave its reluctant recognition to the union.

Some of Local 100's nasty skirmishes belied a creative organizing spark. On September's opening day at the Metropolitan Opera, in the fall of 1999, scores of sequined and tuxedoed ticket holders hoping to hear *Pagliacci* stepped out of their taxis at Lincoln Center to a full-throated chorus of "Met Opera. No more lies. Give us the right to organize." Opera buffs gaped at the Teamsters, brick haulers, seamstresses, professors, singers, and teachers who had gathered to raise their dissonant voices against the Met's subcontractor, Restaurant Associates, for failing to recognize its ninety-five cafeteria workers' desire to have a union. *Pagliacci* went off as rehearsed. Local 100 was on a roll, battling for the low-paid workers who literally fed the benefactors of the nineties' investment boom. There was reason to expect that 60 Wall would follow suit without much resistance.

Brenda's first reaction to the union presence at J. P. Morgan was fear for her job. She had heard about the turmoil over at Smith Barney's cafeteria. The union, to her, meant more raw exposure to management's indignities. Its lead organizer paid a house call one Sunday afternoon trying to elicit Brenda's support. "She yelled at me," Jose Maldonado remembered with a smile. "She told me up and down that she was not joining anything." Brenda told him she didn't entertain unannounced visitors on her weekend. "I didn't appreciate him just stopping by on a Sunday, and I let him know," Brenda said. "I can't afford to go on strike," she said. "Everybody in here would be cut off if I got fired. That job is our bread and butter. I would be straight back on welfare if I lost my job. Meantime, I'm gonna get my respect by myself. That's it."

By September, Brenda was passing out buttons that said "Eat Union Local 100" under an apple with a bite missing. She attended organizing meetings. She worked side by side with the floor manager, Melinda Spooner, talking her coworkers into signing union cards. It was something Jose said that had changed her mind, something about power in numbers. She had realized that Robert's prediction that she would advance up Aramark's ranks was not going to come true. She was better off linking her interests with the whole shop rather than wandering into the boss's office solo.

By the first week of October, all but five of Aramark's seventy employees had signed up with Local 100 for the right to organize. It was a huge victory. Few thought unity was possible in this group. The next challenge was for Aramark to recognize the workers' intent. Six weeks after the dishwashers, grill cooks, cashiers, and maintenance men presented management with their petition, there had still been no response. It was a classic tactic—stall as long as possible, causing defections in the ranks of the skittish. Right before Thanksgiving the union rented a turkey costume, draping its faux-feathered neck with a hand-painted "Aramark" sign. It was meant to be the centerpiece for the union's rally in front of J. P. Morgan's vast granite pillars. Customers had been alerted to the protest via a whispering campaign by workers throughout the bank's headquarters. Brenda was a trusted conduit. Her list of J. P. Morgan friends became an invaluable organizing tool. Several Morgan employees planned to join the picketing. Before the protest could kick off, Aramark relented. J. P. Morgan didn't particularly want the pre-Thanksgiving publicity. The workers at 60 Wall had a union. The next step was forging a contract.

At about the same time a recurring nightmare returned in Brenda's life. A familiar voice called out her name one October afternoon as she walked with Ty in front of Urban Horizons: "Brenda! Brenda!" The sound of that baritone sent her back to the turmoil on Valentine Avenue, just before her odyssey into homelessness. "I was passing by the neighborhood," Brenda would later remember him saying. She turned around slowly to see Teddy Jenkins walking with a faint limp up Gerard Avenue. Brenda signaled to Ty's baby-sitter across the street to get a good look at the big light-skinned black man in the cap. "That's your father," Brenda whispered to Ty. "You don't ever go off with him, you hear me?" Ty, being gregarious and five, rushed into the big man's arms. Teddy told the judge later that his "heart melted" when he saw his son after so many years. He was sick, he said, with hepatitis C. He wanted to spend some time with his son.

Brenda was certain this meeting was a prelude to something more than paternal affection. Maybe Teddy did feel a fatherly urge after five years. But Brenda thought his bigger motive was money. "He hated paying child support directly to me," she believed. And if he had to pay, then he wanted contact with Ty in exchange. Or—her worst fear—maybe he wanted to take Ty permanently. Brenda thought for sure Teddy would have a tough time convincing any judge he was a fit parent. After all, she'd had him arrested for punching fifteen-year-old

Loreal in the jaw, when Ty was a newborn. And he was on parole for life for a rape conviction in his home state of Georgia.

The next month Ty's father reappeared on Gerard and 168th Street. Richard Cuevas was watching his two granddaughters and Ty play on the sidewalk opposite the school when Teddy Jenkins pulled up. Teddy got out of his car and called after Tyjahwon, scooping up the giggling kindergartner into his arms. Richard, the baby-sitter, was terrified. He couldn't physically fight this man if he decided to spirit Ty away. After a long minute Teddy slid the boy back down to the sidewalk. Richard hurried the children up to his apartment and never let them play outside again. The fear of losing her son to this man with a violent past consumed Brenda's days and nights. She filed an order of protection against him. She traveled to family court to petition for legal custody of Ty, a protection she hadn't known she needed until now. Each court date meant a day's pay docked.

Teddy worked up his own homemade custody petition in response. His legal language was delivered in thumping, evangelical tones, honed from years of filing petitions from his jail cell. Brenda, he wrote, gave birth to Ty only so she could "have a meal ticket for the rest of her life." It was he, Teddy wrote, who stayed home from work to prevent her from having an abortion. Teddy accused Brenda of making a living as a prostitute, a topless dancer, and a welfare con artist. "The job she now has is the first legal job she had in many years past," he said.

In a separate letter addressed "Dear Miss Field," Teddy said Brenda's legal maneuvers pushed him from just demanding visitation rights to petitioning for full custody:

> For [I] know now that when I am gone and there is no more money you are going to ditch my son. . . . I want you to know I do not hate you for this, matter of fact, I am praying for you because I know that you have a real problem. . . .

However I am sorry that you could not allow yourself to be a Christian about this and let my son be raised by both his parents. Now I have to think about his well being.

I will see you in court, and I pray that you will try to find some help for yourself. God is the answer to all of our problems and with him all things are possible.

In Christ Jesus,
Teddy

One weekend afternoon before Christmas, Brenda was out of sorts. Still dressed in a leopard nightgown, strands of hair flying like black lightning from her tight knot, she hunched over stacks and boxes of papers, trying to locate the ones that would ensure Ty's custody. His immunization records, his drawings, his schoolwork. Her parent participation certificate from Head Start. Ty's class pictures. "I suppose a mother who doesn't love her son has all these class pictures," she mumbled. Brenda found the criminal court record charging Teddy with assault and endangering the life of a child when he fractured Loreal's jaw. Finally, she put her hands on an inch-thick court document from Macon, Georgia. The legal brief was nearly thirty years old, yellowed, smeared with ink. Touching it again conjured up the terror she had felt when she first found these papers hidden in the back of her closet on Valentine Avenue.

The grisly details of Teddy's past were imbedded in the court transcript. It said that on August 5, 1969, two Georgia friends dropped Teddy off near the home of Mrs. Maureen Hartley, a seventy-three-year-old white woman, and a neighbor of the Jenkins family. Teddy ransacked her home while she slept, looking through drawers, striking matches as he went from room to room. Mrs. Hartley was nearly deaf. It was unlikely she heard the commotion. When Teddy reached the old woman's bedroom, he raped her and beat in her face with an unknown object. Then he dragged her hemorrhaging body from her bedroom, into the hall, out the back, down the stairs, and into the yard where she lay undetected for eleven hours. Mrs. Hartley was paralyzed on her right side as a result. The injuries to one eye left her partially blind. She lived in that condition until she died several years later, never again able to feed, dress, swallow, or bathe by herself.

Police found Teddy's cap under her bed. He pled guilty and was sentenced to life in prison for rape, and to fifty more years for burglary and battery. "Based on what the evidence says," Judge Walter McMillan said at the sentencing, "you have all but taken a life in that the woman that was the victim of your act lays helplessly in a hospital. For all practical purposes she may be living a life worse than death." Teddy served slightly more than twenty years in jail, all the while filing complaints, claiming his guilty plea was coerced and his counsel was inadequate. He won an early release, on lifetime parole.

"He's out to get me now," Brenda said. "But I'm not afraid of him."

In mid-April, Brenda and Ty showed up at Bronx Family Court for the third time on this case. Brenda had all her papers together. This time they were not tossed in a plastic baggy, but rather neatly organized in an accordion folder. The vice president of J. P. Morgan had helped her arrange her documents under categories, like "Ty's Records," "Order of Protection," "Criminal Assault Records," complete with duplicate copies. Teddy walked in late with a member of his new church acting as his counsel. He limped past the bench where Brenda was reading, avoiding eye contact with him. He was wearing a sporty jacket and a baseball cap. Ty looked up from his dinosaur sticker book and whispered in awe: "That's the man who wants to take me from my mommy." Then he ducked playfully under the bench, with a nervous smile. The day ended like so many others, with no decision. Brenda's lawyer was detained in another courtroom. In fact, this case would drag on for another impossible year. Brenda left that day discouraged, making a mental vow to change Ty's last name from Jenkins to Fields, to forever erase this man from her life.

• • •

The court case, the stress, the frustrations on the job had taken their toll on Brenda. The manager asked her to transfer to another cafeteria. She said no, thinking he just wanted her out because the union was negotiating its first contract. She got in further trouble when a couple of cashiers complained that Brenda was "talking union" on the job. Brenda shot back that she wasn't losing her job over gossip and hearsay.

Before it was all over, she checked herself into the hospital. Her arms were hot and tingly. She felt dizzy. Her heart was racing. The nurse in the emergency room measured her blood pressure at 180 over 110. It was in the danger zone for a heart attack or a stroke. (Normal readings for Brenda were closer to around 160 over 80). The doctor put her on a clonidine patch to regulate her blood flow. A conscientious social worker at Montefiore Hospital helped Brenda obtain temporary Medicaid coverage to pay for the hospitalization and the medicine. But the coverage expired after thirty days, just as the patch was causing her skin to burn and her heart to race at night.

She didn't know how much more she could take.

15

Christine

Arrested

It was too hot, and too quiet, to sleep in the shelter. Christine discovered the nights were disturbed by her kids' absence. She cat-napped at night and sleepwalked during the long summer days. At first she tried to make the best of it, using the time without child care worries to find an apartment and get herself back to White Deer Run in the Pennsylvania woods. One solid month in that helpful setting, she imagined, and the black night air and the delicious deer visits, would spirit her cravings away. Magic was all she needed. She was sure of it. It was early June. By September 23, she would get her children back and jump-start her life.

She once had controlled this urge to use heroin, not so long ago, for four years in a row. She could do it again now, on her own. Everyone wanted her to check into a residential drug program, an even more austere choice than White Deer Run. Everyone—Joseph, Wanda, Luis—thought she could use the supervision, at least for a while. That was just like jail, in her opinion. Patients lived in a dorm-like building for six months, the first month in virtual lockdown. She wouldn't do it.

Joseph let her move her belongings into a single room in the shelter, but only for thirty days. That was the state limit for a shelter resident whose children had been removed by the courts. This was a family shelter. She had lost her family. There were hundreds of people sleeping on the Emergency Assistance Unit floor waiting for shelter space, Joseph said. His hands were tied. He could evict her just because she had once been jailed overnight for shoplifting Tylenol from Rite Aid. In later months he could evict her for simply having her welfare case closed. "I'm not doing you any favors to keep you here," Joseph said. "There are drugs everywhere. You need to find an alternative way to live."

The ultimatum was fine with Christine. She would just work faster to find an apartment that would accept Section 8, the federal housing subsidy. She had her priorities down, her route mapped back to respectability. Apartment, detox, job. Apartment, detox, job, children. Apartment, detox, job, children, and maybe Luis back in her life.

But her homemade mantra provided a flimsy defense against the raging urge for heroin. The children weren't there to draw her back into a responsible life. Luis had pushed her out of his life. Nothing else mattered to her. Heroin, the ultimate anesthesia, filled their absence, killed the pain. All it took was one buy, one bag, and the lure of the drug was already more powerful than her will to stop it. I'm already dirty, she would lecture herself. I already screwed up. There's no more "clean" state of mind to preserve. So I may as well keep using. She would snort and feel her stress melt into a haze of well-being. She took some pride in not shooting the stuff into her veins. She wasn't a pockmarked junkie. But when the high wore off, she would feel physically ill. Sober, she could see her complexion turn gray, her bright eyes cloudy. Sobriety brought depression, each one deeper and deeper.

"I'm nothing," she told Luis when he helped her move to the single room. It was a revelation to her, not a cry for help. Her new identity. "I used to be a mother. I used to be a medical assistant. I lost everything. Now I'm nothing. I don't even have a place of my own to live."

Scraping together ten dollars for a hit took over her daily existence. It became her main pursuit. She was giving Luis the bulk of her welfare check to spend on the children. He was caring for the baby and Mark, and getting the girls overnight once in a while. Christine was allowed only supervised visits. She was left with food stamps, which was enough to keep her fed, but not enough to feed her habit.

She turned to stealing. Two near-arrests for ripping off the local Rite Aid pharmacy again convinced her that shoplifting was not the best way to acquire quick cash. Selling drugs was the next option. It was fast. It was good money. She promised herself she would not make it a habit.

Christine approached Robert, one of Cypress Avenue's biggest drug dealers, and asked him for help. Christine was fond of Robert. Even though she never learned his last name, she considered him to be a trusted friend. The drug business was often the only lucrative enterprise on the block. Street-level drug kingpins were often admired as respected citizens in their neighborhoods—heads of the poor person's chamber of commerce. The more simpatico among them felt responsible for the people who worked street shifts for them and lived in their business district. Some dealers threw block parties, others held turkey giveaways on Thanksgiving.

Robert was either one of the generous "entrepreneurs" or a skilled customer sales rep. Christine chose to believe he cared about her as a human being. When she was stuck with no money for the subway, he would slip her a few dollars. Along with handouts, he distributed unsolicited advice. "Christine," he told her just two weeks before the city took her kids, "you are going down. You don't look right. You're asking for trouble. You should get your kids and get outta here. You're going down." She was grateful for the attention. But she ignored the advice, at the time.

For a favor, just this once, Robert gave her ten bags of heroin to peddle on the streets. She could keep one bag and pocket a few dollars from the sale of the rest. But, Robert warned her, today was not a good day. Undercover cops were all over the corner. "Hang with me up the street," he offered. The acupuncture treatment place was serving free meals for Father's Day. It was June 16, 1999.

She didn't want to wait. She was broke. She had an appointment later to sign a lease for a new apartment. Jackson Avenue Family Shelter caseworkers had given Christine a list of rentals that took federal and city housing subsidies. The Coalition for the Homeless showed her where she could get used furniture using subsidized vouchers. She'd never sold drugs before, but she figured she knew what to do. Her younger brother did nothing but sell drugs his whole life. Plus, she'd made enough buys to know what was what.

Christine hung out for a while at 139th and Cypress, a scruffy corner behind the shelter next to a weedy lot, one of the remnants from the time when nearly all the South Bronx was a patch of rubble. Children in P.S. 30 nearby were settling in for lunch. One unkempt man kept approaching people loitering on the corner, asking to buy. No one would sell to him. All the dealers figured him for a cop.

"The guy came up to me," Christine remembered. "I told him no. I didn't have nothing. A guy standing behind him looked all ratty, like a junkie. In fact, I knew he was a junkie. I'd seen him before doing drugs on the corner. While the cop was talking to me, I motioned with my eyes to the other guy to follow me."

Christine took the junkie around the corner. He held out twenty bucks. She slipped two bags in his hand and took the cash. In five seconds two police vans roared out from nowhere. Cops on foot rushed from around the corner. One slammed her body up against the van to frisk her. The junkie was working as an informant. Her bills were marked. She had been caught, on her very first misguided venture.

"Who are you working for? Who are you working for?" the officers yelled. The police weren't after the small-time Christines of the corner. They wanted Robert, the big dealer with the Playboy Bunny tattoo on his neck. She could deliver him.

All the options collided in her mind. She could finger Robert and get a lighter sentence. But he was a friend. He was kind to her. And he was suffering from the later stages of AIDS, fated to die in a few months. If she didn't turn him in, the judge would probably throw the book at her. Then, of course, she wasn't stupid. Robert would have her killed if she squealed. His friendship had its practical limits. "My choice was die now or lose my life in prison."

She told the cops the dope was hers, nobody else's. They charged her with class B felony. "That's B as in Bad," said Rachel Dole, her legal aid attorney known for a battling spirit. The charge carried jail time from two to six years. The judge set bail at $5,000 plus $1,500 cash and sent Christine to Rikers for six days to await a hearing. Hers was known as a school case—conducting a narcotics sale within one thousand feet of a school, in this case P.S. 30, the institution made famous by Jonathan Kozol's most recent book, *Ordinary Resurrections*. It was a serious charge, and very common. "Everything in the South Bronx is a school case," said Dole. "Everywhere is four blocks from a school."

Dole was furious. She claimed the judge was locking every narcotics suspect up that week, including first offenders like Christine. Something was up. It meant Christine couldn't get a treatment deal— drug rehab and probation, with the charges eventually reduced if she cooperated. New York State was known for its stringent "Rockefeller Drug Laws," named for former Governor Nelson Rockefeller, who pushed through some of the nation's harshest drug penalties in the seventies, at the early dawn of the crack cocaine epidemic. Saddled with the stiff charges against Christine, the best Dole figured she could do was to plea for a lighter jail sentence.

When Christine finally was brought in handcuffs before the Bronx narcotics judge one week later, Dole negotiated a higher-stakes deal. Christine would not spend another day in jail as long as she attended drug rehab and stayed clean. If she made any more mistakes, she would be sent to the state penitentiary, not for the maximum two to six, but for three to nine years, at the discretion of the judge. To protect Robert, she agreed to carry this extra threat on her ever-fragile shoulders. Now, instead of teetering on the edge of self-sufficiency as she had planned, Christine was one more mistake away from serving a substantial prison sentence, and losing her children for good.

It was an unusually warm October morning in the South Bronx, the kind of weather that jars the human senses and confuses the flower cycles. Christine rose early. Roosters and salsa music harmonized outside her new apartment window. The No. 6 subway train clattered over Southern Boulevard within shouting distance of her new apartment. Workmen hollered and revved their tractor-trailer engines as they unloaded their cargo of steel beams into the warehouse across the street.

Christine was oblivious to everything but the static in her head. Today was another criminal court date on the drug charge, one of a series to keep tabs on her sobriety. The judge would decide whether she was doing what she needed to do in order to avoid the penitentiary. She needed to concentrate on what to wear. Was this a day for hope, or resignation? Was this an occasion for her black dress and stockings—to show the judge she was together and responsible? Maybe she should just wear layers of sweatpants and shirts. At least then she'd have a collection of practical clothes when he sent her to prison.

She knew she was in trouble. She had tried to follow the plan. Step one appeared to be successful. Christine was out of the shelter and in her own apartment. The apartment was a spacious three-bedroom for $1,050 a month on Lowell Avenue, a short block in the southeast Bronx wedged between the elevated Bruckner Boulevard and a row of apartments on Longfellow Avenue. It was big enough for all her children, once she got her life in order. That was a hopeful sign. Federal Section 8 paid for most of the rent. Her small welfare benefit was docked $110 a month to pay for the rest. Her building stood alone, an urban survivor stripped of all the neighboring apartments that used to butt up against it before they collapsed. Just over the river was Hunts Point and its famous drug spots and streetwalkers—the kind of temptation she did not need. Mark was nearly an adult at eighteen. He moved back in with her. He was working part-time at a construction site learning carpentry and electrical work. He was making attempts to finish his high school degree at Lehman. He dreamed of becoming a herpetologist, inspired by snake handlers in the Bronx Zoo.

Step two, beating heroin, was a different story. She managed to arrange a trip to the bucolic White Deer Run center to clean the most recent river of drugs out of her system. But the experience was disappointing this time. She got sick there. She missed an important family court hearing, the September 23 date that had been etched into her soul last June. Elsie Wood offered blistering testimony against Christine that day, even though Christine wasn't there to hear it. Christine's lawyer, Philip Francis, didn't cross-examine Elsie. He offered no argument in his client's defense, for fear of jeopardizing a later proceeding. He didn't even know where Christine was. She returned home after only two weeks in the woods, without that cathartic feeling of renewal.

When Christine returned from White Deer Run, a Bronx prosecutor then ordered her into a drug rehab treatment near the Cross Bronx Expressway. Christine balked at what she perceived as Project Return's confrontational counseling style. She hated being cooped up for eight hours every day without the use of a phone. Her head pounded with daily migraines, a recurring affliction made worse by detox. Joseph was losing that optimistic note in his voice when he talked about Christine's chances to recover. "She was much more motivated before she lost her kids," he noted. "Now, I can't say what will happen."

Her hold on her three-bedroom apartment was slippery at best. Her welfare case had been cut, again, her check docked. No explanation was offered. The city simply was not paying her landlord the rent money it had pledged. She no longer had Wanda, the Jackson caseworker, to help her figure out things. She couldn't phone the welfare center during the day to find out what had gone awry with her case. So she skipped out on Project Return and headed back for more trouble.

On the first day Christine traveled to her old welfare center in Crotona, the scene of numerous shouting matches, only to be told that clerks had transferred her case to another location. The second day she rode the bus across town to the Willis Avenue welfare office at the Hub—the same center used by Alina and Brenda—to talk to her new caseworker. The third day she discovered the reason her case was closed: "No address." The truth was that, as she moved from shelter to jail to her new apartment, HRA had lost her. She had an address. It just kept changing. This was a common problem for the homeless. She didn't receive the letters that ordered her to come in for a WEP assignment. If she'd gone, she could have argued that she was in court-ordered rehab eight hours a day. Sometimes HRA still counted that as an exemption from WEP. But instead of trying to find her, the city cut her off. As it was, she had few arguments left on her side, and even fewer people patient enough to listen to them. If Christine had lived in any of twenty-six other states, she would have been cut off automatically, and for life. New York was among the minority of states opting not to follow the federal welfare reform law that banned felony drug convicts from ever collecting welfare.

In the course of untangling her welfare case, Christine piled up three unexcused absences from Project Return—just prior to her court date. That was the kind of behavior the judge could cite to send her upstate to the pen. Project Return was a court-ordered alternative to incarceration. She couldn't just skip out. She knew that. Maybe she willed all this on herself. What could be worse than living like this? Then she remembered her six days on Rikers Island awaiting a hearing. The fear, the discomfort.

She stopped a moment in front of her closet, debating her attire, thinking about her girls. They seemed unhappy in their new foster home, during their phone conversations. The city moved Monica and Dyanna from their Bronx home with the pineapple cakes to a permanent foster home deep in Brooklyn. Christine was not allowed to know the address. The new foster mother didn't let the girls use the phone very often. When Christine did see her girls, at first she got disturbing news. Dyanna said Monica was beating her up. She wanted to move somewhere else. Child welfare caseworkers investigated, then decided Dyanna was making up stories about the foster mother leaving her alone, about an older boy sniffing white powder. It was just a seven-year-old's desperate scheme to see her mommy. It broke Christine's heart.

Defying the inevitable, Christine put on the black dress, adjusted her hoop earrings, packed her gray handbag with her journal and her Project Return materials, and traveled on a borrowed token to court.

Streams of sweat trickled down the front of Christine's dress as she climbed the steps of the majestic Bronx County Court building along the Grand Concourse, "the Champs-Élysees" of this struggling borough. The courthouse was built in a pre-Depression era, when residents living along the elegant boulevard had more reason to believe in the promise of prosperity. She felt dwarfed by the austere classical sculptures decorating its facade. She walked around the pink marble figures representing Achievement, Progress, the Law. She passed under the heroic stone figures engaged in work, farming, the arts. Its architects had meant to inspire visitors to the limestone monument with elegant testaments to the work ethic. Maybe they had succeeded in another, more equitable economic era. That day, in October, for Christine, the building just inspired dread. All those statues were mocking her.

Couples dressed in pastel chiffons and tailored suits whisked past Christine on the stairs. The marriage bureau was in the basement. Television camera crews tested their equipment on the court platform. The Reverend Al Sharpton was expected any minute to lead a protest against the court's decision to move the Amadou Diallo shooting trial from the Bronx to Albany. Four white police officers had killed the unarmed West African immigrant in a hail of forty-one bullets eight months earlier, in February 1999. The cops mistook his wallet for a gun. Since then, an anti-cop furor had swept throughout the Bronx and the city. Now the judge had ruled that there were not twelve jurors in the whole of the Bronx who would be unbiased enough to hear this case.

Christine didn't have the luxury to ponder the problems of her community. She scanned for her name posted outside Judge John Collins's courtroom. He was the borough's chief justice. She poked her head in the door and looked around, tentatively, like an errant schoolgirl afraid to face an angry teacher. His courtroom was one of the building's most elaborate. She couldn't bring herself to enter. The room, paneled in intricate carvings from floor to ceiling, bore hints of its grander days. Dusty chandeliers with chains and simple dark brass fixtures hung from the ceiling. Mustard-colored venetian blinds flapped down over the windows, askew, tangled in the middle. The seating area resembled a vandalized cemetery. Many of the scrolled-arm chairs with brown vinyl seats were missing their arms, leaving iron spokes and yellow police tape where the chairs had once been.

Attorney Rachel Dole couldn't make the hearing, but she was not optimistic about the judge's imminent decision. "Ms. Rivera is an addict, and she's acting like an addict," Rachel told an acquaintance, referring to Christine's erratic attendance at Project Return. "I know, she's got a disease. But I'm pissed. The judge is pissed. And the Project Return people are pissed."

Christine peeked through the door three more times. Finally, the court clerk called out her name. She stood at the wooden gate before the judge. Christine shifted her weight from leg to leg while she waited for the judge to finish reading her files.

From Christine's vantage point, Judge Collins was a distinguished, white-haired head, disembodied, poking up from behind a grand, carved bench. He read her documents with studied deliberation, barely flinching in his bright red chair. After a long minute the judge looked over his glasses and directed his gaze at Christine. "Ms. Rivera, you are not doing well in this program," Judge Collins began, his voice gaining in volume. "It's important that you do well. We don't want you to fail. We want you to succeed. If you're not able to lead a drug-free life, all promises are off. I could remand you upstate today for your full sentence. You pled guilty to a class B felony."

Christine fidgeted with her hands, flipping them in and out of her pockets, dribbling some of their contents on the floor. She felt only her own fright.

"I won't," the judge continued. "I will remand you instead to thirty days. We will see whether when you get out you can get your life together. A lot depends on you. You need to help yourself."

A month in Rikers Island. She looked back over her shoulder as two guards escorted her to the holding cell behind the judge's chambers. Her eyes darted over the audience. She was hoping Luis had come. Or maybe Mark. No one was there to say good-bye. She would miss her visit with the girls that weekend. Mark would be alone in the apartment. The black dress and hoop earrings were an ill-fated call.

• • •

ikers Island was a sizable piece of land wedged between Hunts Point in the Bronx and La Guardia Airport in Queens, in the center of the East River. The only way to get there was over a bridge that looked like any other city street, only jutting into the East River. It ended at a jail checkpoint and a parking lot. A city bus took visitors the last mile. Once a landfill in the 1800s, the island became home to the city's massive jail system after the turn of the century. During the Civil War the island was the site of a unique tale in American racial history, when poor urban blacks were offered not welfare, but ammunition. In mid-1863, New York City erupted into violent draft riots. Mobs of mostly Irish workingmen protested the nation's new conscription laws by storming through the streets, terrorizing the wealthy, and lynching black men and women, burning the corpses, and virtually emptying the downtown waterfront of African freed slaves. In retaliation for the riots, according to the authors of *Gotham: A History of New York City to 1898*, "Republicans gave blacks not alms but guns." Republican leaders armed a thousand-man squadron of African American soldiers and headquartered it on the local landfill, Rikers Island. A year later the regiment marched into Union Square in a show of strength against the white immigrant laborers and longshoremen.

More than a century later the city used this once revolutionary training ground to imprison mostly black and Hispanic men and women. Of the 130,000 inmates in any given year, 57 percent were black, 35 percent Hispanic, 6 percent white, and the rest a smattering of Asian and other. But the truly staggering statistic was the surge of women in jail, most of them convicted of drug charges. More poor women, and more Hispanic women, were getting caught for selling drugs than at any other time in our nation's history. Nationally, the number went up 888 percent over a ten-year period beginning in the late eighties. In the city more than half the 1999 female inmate population was busted for drug offenses. Bronx women and Hispanic women were greatly overrepresented in this population. Forty-four percent of female drug offenders were Latinas, even though they made up only 14 percent of the population. Authors of a study on women and drugs by the Sentencing Project theorized there were two strong reasons for the sudden surge in female drug convicts. Women were more reluctant than men to turn in their dealers, so prosecutors were disinclined to offer nonjail deals. Also, poor women were poorer now than they had been five years before. The small welfare benefits, the frequent cutoffs, and fewer livable-wage jobs had taken their toll.

Christine found herself one of the unlucky statistics in the fall of 1999. While Brenda was spending her days running the cash register and fighting for a union, Christine spent her days in a gray prison pantsuit, working the prison bakery ("all the bread you can eat," she lamented), and attending a drug program. Christine could usually find something to be grateful for, a simple mind game that gave her comfort. In Rikers, she was grateful to be an addict. It entitled her to this special program—one she liked, one that motivated her, one that removed her from the more dangerous "population." Meanwhile, her welfare case was still closed. Mark was getting by with temporary help from his dead father's brother. Christine worried about that, and about her girls. When she wasn't working, she was writing letters.

One she left unsent in her spiral notebook:

Dear Monica,

I don't know where to start since you know you can't come home yet. But you know I'm trying. Someday when you're older you'll understand why I'm such a messed-up mother, or why they took you in the first place. I have a big problem. I cannot deal with life on life's terms. If I don't get you back, or I die, I hope you don't hate me. I know I haven't been the best mother. I should've stopped and stayed stopped. It is very hard. It doesn't mean I don't love you. I have an incurable disease called addiction. I hope you never get it, it's horrible. It makes you do things you probably wouldn't do sober.

Please promise me that you won't do drugs and you will take care of your sister and keep contact with your brothers. I hope one day I can look back and say I won't ever let anyone separate us ever again and that you and me can finally get along. If it doesn't happen, I hope life gives you everything it has to offer. Never settle for less. You're just as good as everyone else and you're a beautiful young lady.

I'm sorry if I cannot be there when you have your first boyfriend, or your wedding, or your first child (if you have any). Please look out for your little sister, you're basically all she has. Remember the misery and unhappiness, the loneliness I had to go through with drugs and alcohol and where it got me. Nowhere. And so miserable and couldn't stand to be with myself.

Love always,
Mom

A few weeks later Christine was out of jail, back on heroin, and back in rehab. The staff checked her into the hospital psychiatric ward for suicide watch.

Part IV

Hearts to God

The test of our progress is not whether we add more to the abundance of those who have much: it is whether we provide enough for those who have too little.

—FRANKLIN DELANO ROOSEVELT

16

Case Closed

Three days before the millennium, reporters gathered in City Hall to hear the mayor's final report card on his own maxims for the poor. Two years earlier he'd made some particularly bold prescriptions. Come 2000, the mayor had announced, methadone treatment for poor heroin addicts would no longer be sanctioned with city dollars. Work, he had said, combined with other kinds of non-opiate treatment, was the healthier path for substance abusers on welfare.

But even more monumental, Giuliani had declared that welfare would end, not just for addicts and the homeless but for everyone. The culture of dependency, propped up by liberals still trapped in the "poverty industry," would be replaced by a culture of self-sufficiency.

His self-made deadline was now just days away. Giuliani took the podium to tell the city how he had fared. He didn't mention methadone. That issue had long since become moot. The mayor came to realize he had regulatory power over only two thousand addicts in city hospitals.

The other thirty-six thousand were state and federal patients, out of his jurisdiction and control. Giuliani barely mentioned the hoped-for conversions of all income support centers to job centers. At that time those plans were still stalled in federal court. Judge Pauley was reviewing rooms full of legal aid data that purportedly supported the argument that the city was still turning away the hungry and the desperate.

Undaunted by these setbacks, the mayor declared victory. The millennium, the mayor said, "marks the milestone of replacing the culture of dependency in New York City with the culture of work." Nearly half a million people had disappeared from the rolls. Those still unfortunate enough to be shackled to a benefit check were all working for it, he said.

The *New York Daily News* declared in its headlines the next day that welfare was dead. What did that mean exactly? HRA was not closing up shop. The city was not going to cease cutting benefit checks or dispensing food stamps. No one was arguing that poverty had disappeared.

The truth lay somewhere in the semantics of the moment. After Giuliani declared his intention of ending welfare, he and his welfare commissioner spent the next two years refining what ending welfare meant. Both men said they never vowed that everyone would be working for living wages by the year 2000. Instead of "work," the two men preferred the phrase "engaged in a work activity." Their goal, they said now, was to engage every welfare participant in something that simulated a thirty-five-hour workweek. Four months before the millennium, Turner still needed to find something for forty-eight thousand more recipients to do, by his own estimate, in time for Giuliani's deadline. It could be simulated work, in a way. It could mean a drug treatment program, a job search, a WEP assignment. It could even mean simply that the client received a city letter outlining the work required of her. One HRA worker joked that recipients might be ordered to ride the subways and read posters for thirty-five hours a week, just so the mayor would be able to greet the new year saying no New Yorker was receiving something for nothing.

The real numbers depended rather specifically on who was counting whom. Months after the mayor's announcement, the city's Independent Budget Office found evidence that the end of welfare was little more than a wistful illusion. The analyst Paul Lopatto reported that only 29 percent of the total 267,000 welfare cases were "fully engaged." That number included 36,000 WEP workers, others in training programs, and teenagers in high school. More than 34,000 others were waiting for administrative judges to decide whether their benefits should be restored. Most of those cases would be decided in favor of the recipient. About 23,000 others, uncounted, were enduring sanctions for disobeying the rules. Ending welfare, it seemed, was more difficult than shuffling numbers.

Still, when Mayor Giuliani stood before the press on December 28, 1999, he claimed that the number of idle people accepting city gifts and giving nothing in return was now absolutely zero. If all hearts weren't with God, then at least all hands were allegedly busy.

Christine Rivera walked across the span of the South Bronx the day the mayor declared that welfare had died. She traveled these nearly three miles every day, from the Bronx River on the east to the Hudson River on the west, to her court-appointed rehab. She didn't have the three-dollar round-trip bus fare. Christine was one of the thousands who did not figure in the mayor's tally. Her welfare case had been cut off a month earlier, for the third time. The reason given: no address. She was sick and broke and barely participating in her own rehabilitation.

So Christine braved the December 1999 chill in her black, puffy jacket, crossed under the No. 6 elevated tracks over Jerome Avenue, passed the auto chop shops, the ninety-nine-cent stores, to Project Return, an alternative to incarceration. She was late, again, something that caused her head to throb. This could cost her again. She was on a "last chance" contract with the drug rehab program. The counselors were doing their best to keep her on track. Still, one more mistake and she wouldn't see her kids for years. Christine buried her nose inside her jacket and kept moving.

Christine's already precarious life took a deeper plunge. Project Return staffers recommended that she enter residential rehab, the program she had avoided like the plague in the past. The streets still had a damaging hold on her. She needed stricter supervision to conquer her addiction. But there wasn't space soon enough for her, not just yet. In the meantime Christine slipped into a deadly pattern. She missed days and weeks of rehab as her heroin habit completely consumed her. Christine would check herself into a hospital for detox, then slip right back out to a dime bag and a sick high. By February she was turning tricks along Westchester Avenue, giving some of the cash to her oldest son, and smoking crack with the rest. "I hated it so much," she would say later. "But the money was amazing." One Project Return counselor shuddered: "She looks like she's dying."

After one four-day binge, Christine found herself gray, mottled, sick, and crying on her hands and knees before the Samaritan Village rehab center in Briarwood, Queens. There wasn't room for her in the Bronx. She begged the staff there to take her in. One more day on the street and she knew she would die. After a brief assessment, Christine boarded a bus up to the Catskills where Samaritan Village ran a lockdown facility for 240 substance abusers. Evaluators assigned her to twelve to eighteen months of hard-core treatment at the Ellenville, New York, campus, where counselors used holistic techniques to break down their patients' deadly habits. Welfare kicked back in to pay for the costs.

Christine was back in the woods, whose peaceful calm had almost cured her years before. There was no Cypress Avenue street corner to tempt her habit, no stores to rob, no way to escape, except on foot. Her urge for a heroin high melted away as the stars grew closer and the owls hooted louder. After seven months, Christine showed up in Judge Collins's elegant courtroom looking sophisticated and sober. Her often unruly hair was cut in a stylish bob of waves. She wore a tasteful long knit skirt, black blouse, and sandals. Program counselors promoted her to the second level, giving her more freedom and more responsibility. Christine stayed on track, and she graduated to a Manhattan halfway house by New Year's 2001. By then, her lifetime welfare deadline was depleted by three years. Two more years to go. It was enough for Christine to imagine embracing her mantra again. She would get her apartment back, her kids returned. She enrolled in a Manhattan school that taught ultrasound and radiology techniques. She would get a job at Bellevue Hospital. She would have Luis back in her life. Maybe then, she could have a fifth, or sixth, or maybe tenth chance to get it right.

Alina Zukina was finishing up her first-semester final exams in biochemistry and anatomy at the New York College of Osteopathic Medicine in Old Westbury, Long Island, when the mayor declared that welfare was at an end. Perched in front of a jury-rigged desk at her Bayville window, looking out over the Atlantic surf pounding the shore, Alina could barely summon up the dreary memory of an income support center. Her case had been closed for good a few months earlier.

For a landlocked Moldovan, raised to ration bits of coal, this new existence represented a briny freedom. Alina was living on her own for the first time in a way. An Indian, a Syrian, and another Russian student named Marina shared this beachfront rental with her. Alina's mother paid the $470-a-month rent out of her home attendant wages.

Lyusya also managed to feed her skinny daughter every meal but breakfast by shuttling a two-week supply of lunches and dinners from the Bronx to her Long Island door. Lyusya filled up fourteen Chinese-food plastic containers with different soups—borscht, chicken noodle—Alina's elixir. Fourteen more contained little meals of stuffed cabbage, chicken and rice, and mashed potatoes. She then paid the man who delivered Russian bread to the local Bronx grocery to drive this carefully parceled collection of food relief to Marina's parents in distant Brooklyn. There, this other mom and dad would add their own fourteen-day supply of food to sustain their daughter, and then drive the remaining few miles to deliver these catered goods to the quaint village of Bayville. When the car arrived from its bimonthly trek, the house would break out in a whoop: "Here come the meals on wheels! Soup kitchen! Soup kitchen!" Alina's mother put her twenty-five-year-old daughter on the scales whenever she returned for a weekend visit to see whether this production-line service was providing sufficient nutrition.

Medical school was stimulating, but stressful. "I almost quit in the beginning," Alina said, recalling the overwhelming schedule. "We do in two years here what they study six years in Kishinev. It was too much stress." Time soon soothed her nerves, and good grades rebuilt her confidence. She fell into a rhythm. She found her round-the-clock study pace. It helped that about 30 of the 240 students in this branch of New York's Institute of Technology were Russian immigrants too. "I'm happy," Alina said, looking out over her new quaint town, wrapped gracefully around the bay. "I'm doing what I want to be doing."

Alina traded in life on the dole for life in deep debt. Federal loans paid the school's pricey $24,000-a-year tuition. Alina had very few other expenses. She used the college's computers. She rented a microscope. She borrowed her roommate's stuffed monkey to practice osteopathic techniques. She dropped a few dollars a week on breakfast, for cottage cheese and sour cream. The loan was gathering interest at a rapid rate. She would have to pay it back quickly after graduation. But for the time being that didn't worry her. Alina had a histology test to take, a pharmacology lecture to attend, a future to plan as she inched ever closer to her dream of becoming a family practice physician. She and her Moldovan boyfriend, Mark, were making plans to marry. Welfare would never be an option, nor hopefully a necessity, for her again. She had used up her lifetime limit.

Brenda Fields was in the middle of an especially long week the day of the mayor's triumphant announcement. She agreed to work the New Year's Eve overnight shift on Wall Street for time-and-a-half wages. Brenda signed up to serve food to the thousands of J. P. Morgan employees who were working on an emergency basis to thwart the feared millennium meltdown. The apocalypse never materialized, of course. Instead, New Year's Eve turned into one of her last stress-free nights on the job.

A few months later, on a Monday in early spring, Brenda awakened for work at her usual hour. She left Ty with a security guard at school before the breakfast program opened up so she could make it to work by 8:00 A.M. She'd missed a lot of work since January with her hypertension and with all her custody problems.

Each day she missed was a day's wages lost. She sprinted down Wall Street, passed between J. P. Morgan's pillars, walked through the maroon-and-white-tiled kitchen, past the salad bar, the fruit baskets, the gourmet coffee station, to punch in on time. Her day was routine—on the quiet side, in fact. The women Brenda normally rubbed the wrong way were subdued in the busy basement. A rumor had wafted through the cafeteria the previous Friday that the manager had compiled a list of a dozen people he was about to fire. Brenda's name was one of those in the whisper mill. But she put it from her mind. She decided instead to seize the temporary serenity and enjoy her day. Several J. P. Morgan customers visited. She was a good listener. Employees often came down after the lunch rush just to chat with Brenda about their work, their boyfriends, their mothers and sisters. They were all ages, all races, all demographics. Asian, white, black, young, old, secretaries, executives. Brenda joked that she was going to start charging a multicultural therapy fee one of these days.

She was fifteen minutes away from ending her shift that afternoon when her boss appeared. "I need to talk to you," he said. "Just you and me." She was cleaning up the Java City coffee station with Melinda by her side. Brenda looked up at Melinda. As shop steward, Melinda was usually invited when employees were in trouble. Brenda followed Owen to his office alone. The general manager told her he had reason to believe she had taken "product" from the cafeteria last week. He had witnesses. He had proof. He told her it was J. P. Morgan's "product," and he couldn't have it leave the building. He asked for her time card and employee identification. Brenda burned inside with this accusation. He was calling her a thief. It mattered to her what the J. P. Morgan employees thought about her. She wouldn't be around to defend herself. But all she said was, "Oh, okay. Do you need anything else? My uniforms?" She was banned from the building, stripped of her identification. She became an official security risk.

Brenda walked out of Aramark's cafeteria that day feeling an odd sense of relief at first. She wouldn't have to spend her days with people who hated her anymore. She could pick up Ty every day from school, pay attention to her health, be rid of all those back-stabbing gossipers. She could shore up her family. Maybe she was free to find herself a better job with a future.

But as the days progressed and the facts unfolded, her short-lived relief turned to lingering indignation. The accusation became more precise: J. P. Morgan surveillance footage showed Brenda leaving 60 Wall carrying a small takeout box, presumably filled with leftovers.

Security did not stop to search her. Brenda was being fired for taking out a box. Aramark had once trusted her with the keys to the entire cafeteria. Now she was a criminal, unfit to visit her old friends. It was an undignified way to end eighteen months on a job she once cared about. Dozens of Morgan employees called her home, leaving messages of support.

When union leaders showed up at Owen's office to view the evidence, the general manager presented only one employee, Melinda Spooner, the shop steward. Melinda told Jose Maldonado, Local 100's lead organizer, that she saw Brenda standing in front of a salad bar, holding a takeout box and a bottle of water. She said she didn't see Brenda put anything in the box. Nor did Melinda see Brenda leave the building with it. Brenda told Jose she often put her extra shoes or customers' gifts in those boxes. It wasn't necessarily filled with Aramark's leftovers. Jose was furious with Melinda for placing the union in this untenable bind. Local 100 was still in negotiations with 60 Wall for its first contract. This was the time management often seized to force out unwanted, politically active workers before an official arbitration process was in place, Jose said. "You know your job is not to turn coworkers in to the boss?" he seethed at Melinda. "Your job is to unify your workers." Melinda burst into tears, saying she feared for her own job.

Within one week Brenda's indignation turned to gnawing despair. Freedom from the injustices of the workplace carried a steep price. She was completely broke. She spent her last paycheck on groceries. She delivered what was left of her cash to Con Edison to make sure the utility didn't turn off her lights. "These kids are not going to live in the dark," she said. After that, there was nothing left to cover April's rent. Brenda cried with the same despair she had felt during her darkest hours looking for this job. "I feel I've let down these kids in here. I failed them. I should have held out for a better job. I should have gotten my education. I'm not sure what I did to deserve this."

She spent one day at the unemployment office, figuring out how to apply. After some arm-twisting by the union, Owen promised he would not contest her appeal for unemployment. It would take more than a month before she saw a check. In the meantime Brenda traveled to the Highbridge Food Pantry for emergency supplies. WHEDCO's social worker offered emergency soap and toilet paper, cash, and a temporary rent waiver. Her heart raced at night. Her head was foggy during the day. She was short of breath. Her skin was burning from the clonidine blood pressure patches. She spent a third day at Montefiore Hospital getting a new prescription, then traveling from floor to floor hoping to find a way to pay for it. She had no health insurance. She had a Metrocard in her pocket, but no money. After four hours she got the fee waived—but just this once. Brenda was running out of choices. The most disagreeable option, welfare, seemed unavoidable.

On April 10, 2000, the *Daily News* lead editorial declared that welfare was New York's premier success story. "For the first time in more than three decades, the city's welfare rolls have dropped below 600,000," it read. "That's not only historic, it's a tribute to Mayor Giuliani's steadfast determination to break the cycle of dependency for hundreds of thousands of poor New Yorkers." There was no mention of how many of the former dependents were making a livable wage. Nor how many were forced to return to the rolls when the labor market could no longer sustain them. After four years of welfare reform, those numbers were tough to come by and did not make for a tidy sound bite. "This is all by way of saying that the new New York century is off to one helluva great start," the editorial continued. "And everyone benefits."

The editorial ran the same day Brenda went back to the welfare office to seek help. Until she stepped back into line, she was included by implication in the editorial writer's list of statistical successes, supposedly one of the hundreds of thousands who had benefited by dropping off the rolls. In fact, she was considered one of the system's proudest stories. Brenda held a corporate job for nearly two years. It was steady, it offered health insurance, it was becoming a union shop. Her salary was even above the poverty level, if only by a notch.

Brenda caught the city bus to 161st Street and waited with mobs of people for the elevators to take them up five flights to the Melrose Income Support Center. Brenda had less time now than before to get lucky with a decent job. She'd used up nearly three years of her five-year lifetime limit. Nothing much had changed since she was here last, in the winter of 1997. Melrose was one of the fifteen centers that had never been converted to a job center. The scruffy waiting room was packed with myriad stories and raw tempers. The blue plastic chairs couldn't hold the crowd. The staff was overworked and inefficient. Babies were crying to her left, to her right. She filled out a book of forms. The workers fingerprinted Brenda, again. They took her photo, again. They gave her three more appointments, again. One was back at social services, one in Brooklyn's fraud bureau, one in Manhattan to pick up her photo identification. That would just be round one.

After ten hours of waiting, the clerk offered Brenda thirty-nine dollars in emergency cash. Days later she found out the city had refused to authorize emergency food stamps. The reason: she *might* start receiving unemployment in a few weeks. Instead, her first check was delayed until the following month. Until then, thirty-nine dollars was supposed to keep three people fed, clothed, sheltered, and transported in New York City.

Like about 30 percent of the former welfare recipients in the nation, Brenda was back on the cycle of dependency. It was the very last place she wanted to be. This time, though, Brenda was sicker and more demoralized than she was before. She thought she had done everything according to the rules, new and old. She would figure out what happened. She would summon up the courage she knew somehow was the legacy left by her grandmother. She would not give up.

She believed in work. It was supposed to save her. Instead, she would have to save herself.

Part V

The New American Poor

It's time to honor and reward people who work hard and play by the rules. . . . No one who works full-time and has children should be poor anymore.

—BILL CLINTON AND AL GORE, *Putting People First: How We Can All Change America* (NEW YORK, 1992)

Epilogue

When the nation entered into a vast welfare experiment in 1996, it was divided on many details, but not on the need for reform. Welfare as we knew it was deeply flawed and wildly unpopular. Some thought it was dangerously generous; others found it perilously stingy. Some thought it actually caused poverty; others believed it was a useless and costly barnacle. Nobody believed it was working well to place poor Americans on their feet. Nearly thirteen million people were attempting to live on less-than-subsistence public aid in the world's richest nation—and the numbers were growing. Something had to change.

Nor was the power of work a debatable topic in the mid-nineties. Few on either side of the partisan divide could argue that a job was not the best anti-poverty measure. Work was not a ragged remnant of Puritan times. It was a proven reality. Children whose parents moved off welfare into decent jobs tended to be healthier, more successful in school, and more likely to take advantage of higher education later on in life. The key was a "decent" job—one that provided health care and an above-poverty salary, one that boosted self-esteem, creating a sense of accomplishment rather than a dread of dependency.

By the turn of the twenty-first century, many signs pointed to a hopeful future. Welfare cases continued to drop to historic lows. More than five million people left the benefit rolls after President Clinton signed the bill to end welfare as we knew it. Quality of life improved, in general, across the racial and economic divides. Children were healthier than they had been a decade earlier. Teen pregnancy was still higher than in most developing countries, but on the decline.

America had never been wealthier in all its history when 1999 became 2000. Its economic indicators soared off the global financial charts, while across the country, crime, unemployment, and poverty rates plunged to record lows. More people were working than just five years earlier, and many of them were single mothers. Average salaries rose at their fastest rate in twenty-seven years. Workers on the lowest end of the wage scale saw real improvement in the value of their paycheck—in large part due to a generous federal tax credit for the poorest workers.

Some of the nation's good news was due to the unflagging financial boom of the nineties. Some could be directly attributed to the new policies. But the upward trends told only a fragment of the story. It did not begin to explain the disparate experiences of Brenda, Alina, Christine, or thousands of other individuals who struggled against the rules to make ends meet. It did not explain why there were still thirteen million children living in poverty nationwide by the end of the century. It did not account for the chasm between the very rich and the very poor, or the growing number of homeless, or medically uninsured. It offered few answers for the thousands of the most vulnerable Americans—those damaged by substance abuse, sexual assault, mental illness—who still would require help once their lifetime limit had expired.

Finally, and probably most troubling, this positive news did not explain why a growing number of Americans were working full-time and were still poor. This development had broken the social contract between government and those in need. Most Americans believed that the poor should work in exchange for benefits. Few, however, thought that any parent who followed the rules, left the rolls, and got a job should still have trouble feeding her children. Columbia University's National Center for Children in Poverty reported in 1999 that nearly two out of three poor children lived in a home where at least one parent worked, a figure higher than at any time in more than two decades. These were the new poor in America—the working poor—a startling anomaly in an era of prosperity.

Beyond the initial good news, then, lay the harder questions. Were the poor better off now than they had been before reform? Was our nation focused on eliminating poverty, or just slashing caseloads? Did "work first" really work? Did it make children's lives better, or more fractured? What would happen to those left at the bottom of the economic scale once prosperity embarked on its inevitable slide?

Some answers were found in the reams of statistics, reports, and analyses available to us. But the most honest understanding was still burrowed in the vast complexity of poor people's lives.

The tangled lives of Alina, Brenda, and Christine can be viewed in many ways to justify one political view of welfare or another. Some may say these women demonstrate as a whole that reform is on the right track. Alina might have remained shackled to welfare if she hadn't been forced by the rules to obtain her own college loans on the way to becoming a doctor. Brenda might not have found employment as quickly as she did, gaining valuable work experience, if the deadline hadn't lit her fire. Christine might have used her welfare check to subsidize her heroin habit for a lifetime. Getting off welfare was the first step toward personal freedom.

Still others may say Alina would have had her medical degree years earlier if the city hadn't cut her off welfare so many times on technicalities. Brenda would not be struggling on welfare once again if the system had subsidized her paycheck and provided more generously for child care and health insurance. Ty would not have suffered through a series of traumatic baby-sitting experiences. Christine would not have succumbed to shoplifting and jail if HRA hadn't cut off her check as often as it did. She might not have lost her kids.

If there's anything to learn from these women, it's that partisan platitudes about them are fairly meaningless. Their lives are nuanced and messy. These are three women living in the same neighborhoods, in the same borough, in the same era, and yet they are each unique, each complex, each challenged in different ways. One desires a livable wage, another addiction therapy. One wants a college degree, another needs to settle her custody battle. One may require strict rules to get her out of the house; another may teeter on the brink of death because of them. All three women could use a more equitable job market, higher wages, and better child care. Rules and deadlines meant to urge individuals into the workforce failed to address these more sweeping policy needs.

The most telling, predictive factors of success were established long before any of these women stood on a welfare line. A strong education and a reliable family network help smooth the path out of poverty. Labor statistics show clearly that a higher education degree lifts wages by as much as 30 percent. Welfare checks alone rarely provide enough to sustain a family. As Rutgers University Professor Kathryn Edin revealed in her groundbreaking book *Making Ends Meet*, nearly every welfare recipient needs to turn elsewhere for financial help simply to keep her income near the poverty level. That includes those living on New York's welfare grant, which at $577 a month for a three-member family is considered one of the nation's most generous.

Alina arrived in America with no more money in her suitcase than Brenda or Christine carried in their pockets. She felt like a fish in the desert, forced out of her cultural element and into an adopted country where everything from language to pizza was foreign to her. Alina did, however, have solid family ties. Two uncles had already settled their families in New Jersey and the Bronx. Just as Giuliani's father leaned on his family when he lost his job, the Zukinas depended on relatives for years to subsidize their rent and provide psychological sustenance and an instant community. Such generosity was expected, even if the uncles had little money to spare. They were family. Alina arrived at the welfare line already buoyed by family, by race, by the well-funded network of Jewish refugee aid networks, and by a stellar high school education. Her schooling in the Soviet Union was so rigorous that it prepared her to make straight As in an American college. When she finally made it to medical school, her mother, using an elaborate network of friends and bakery men, delivered "meals on wheels" to her at her Long Island college.

With more resources to draw upon, Alina managed to duck under the restrictive WEP requirements just long enough to qualify for medical school. She survived, without a ruble to spare. Navigating the system, learning its rules and quirks, transformed the shy Alina into a tough New Yorker in spite of herself. I was often embarrassed when I witnessed how rude some welfare workers behaved toward Alina. While a few sympathetic officials whispered to her that she should pursue her degree until the system caught up with her, others lectured her on the value of a job, insinuating that she was bilking American taxpayers to get ahead. When she was bounced off welfare not once but twice for a paperwork error, Alina simply shrugged. "I will talk back to the welfare," she laughed.

Alina's experience showed how the most well-meaning policies can become absurd and counterproductive when translated into an individual's life. Alina came to America with little except the desire to become a doctor. She embodied the American dream. She could eventually earn a big salary and pay substantial taxes, while doing highly valued work, if only she were allowed to finish her studies. Yet she couldn't complete her studies without help for rent and food money from public assistance. The new welfare world no longer valued a degree as much as it worshiped WEP. The idea was to alter individuals' behavior, not simply hand them a diploma. A great many reformers believed that welfare should not be used as a college scholarship program. The mayor promoted "work first, education second." Learning the work ethic was more important than learning anatomy.

Alina kept her nose in her textbooks, knowing instinctively that the system was upside down, at least for her. She dodged some WEP assignments, contested others, and scooted under HRA's radar until she no longer needed help from the agency. Ironically, it was her work ethic that allowed her to earn straight As in difficult courses in a language not her own. Higher education has always provided a time-honored path out of poverty, particularly for new immigrants and their children. A recent study of five hundred state college students on welfare found that 87 percent graduated into steady jobs, earning upward of $20,000 a year. Researchers have established, not surprisingly, that those with higher skills and better education fare better in the workplace.

By March 2001, Alina was one year away from beginning her residency, and three months away from her wedding to Mark Filizov, her Moldovan boyfriend. She had more than survived the system. She had thrived, in spite of it.

• • •

Brenda's family, in contrast, was fractured from birth. She was raised by a series of foster families, promised adoption that never happened, split from her siblings at age fifteen. Brenda learned early on to adopt her own universe of relatives, in a uniquely creative fashion. An especially close friend would become an "aunt" or a "godparent" to her children. Her daughter's best friends would become "nieces." These were terms of endearment, not bloodline titles. Brenda cultivated her own communities, sharing her home, her food, and any money she might have with neighbors in need. Her friends would often reciprocate when they could. One neighbor turned over her entire monthly welfare check after Brenda lost her job. The service providers in her building, Urban Horizons, provided a surrogate family for her when the bottom dropped out. But in the end there was no real family to count on, unconditionally. She depended completely on the vagaries of public aid in an unforgiving city.

As a result, Brenda's journey through the maze of welfare rules was slightly more desperate. Unlike Alina, she was already a hardscrabble New Yorker. That was not something she needed to learn. Working the strip clubs and catering halls, raising two children, and fending off a heavy-fisted boyfriend, Brenda had earned her street stripes. But she was worn down. The new message of welfare seemed to fit the bill. If welfare was all about jobs, she figured, then there must be jobs to be had. Brenda was hungry for one of them. She had faith in the system. She wanted secure employment with health insurance and a chance for promotion. Brenda followed the rules—or at least, most of them. She made her appointments. She signed up for a job-matching program that gave her one-month grace from the WEP requirements. When she didn't land a job in the thirty days allowed, she hit her first wall. Welfare cut her check almost in half and eliminated help with child care.

Such cuts were designed to force people into the job market by squeezing them financially. But for Brenda the cuts made it nearly impossible to continue her job search. If she took a WEP job, she would have no time to look for a real one. Yet if she didn't take a WEP job, she couldn't afford child care for Ty, or subway fare for herself. Brenda resorted to begging for cheap baby-sitters so she could look for a job. And Ty paid a price. Still, Brenda had the strength of character to persevere. She never stopped believing that in spite of the new system, she would make it.

For a brief while the promise of welfare reform looked real to Brenda. She was never happier than the day Aramark hired her to work in J. P. Morgan's cafeteria. It was a good job, by all accounts. The pay put Brenda above the federally recognized poverty level. Work conditions were decent, until they weren't. The late shift wore her down. Ty was frequently sick from being roused every night at the baby-sitter's so he could walk to his apartment across the street. A promised promotion disintegrated. Still, the strain would have been worth it if she had been making a living wage. At $8.25 an hour, with a family of three in New York City, she was still struggling. The phone, rent, electricity, food, and transportation added up to about $1,150 every month. Brenda was bringing in $1,100 after taxes. There was nothing left for school supplies or clothes. She didn't own a car. She didn't own a computer. She shopped in the thrift stores. When a new manager took over, it was clear to Brenda the balance sheet was never going to improve in her favor. She became increasingly frazzled, missing Ty's school performances, scrimping to pay the bills.

Brenda's story was most directly the story of the working poor, as defined by welfare reforms in America. She couldn't live off welfare. Nor could she get ahead on her paycheck. The Clinton administration tried to ease the squeeze by greatly increasing the tax returns for those earning at the bottom rung. The Earned Income Tax Credit was arguably the single most effective anti-poverty program to emerge from Washington. But it wasn't enough to push Brenda over the top.

Despite all her verve and heartfelt effort, Brenda found herself perpetually jogging through tar.

When Aramark let Brenda go in the spring of 2000 for allegedly taking a salad and a bottle of water from the cafeteria (charges that were later expunged from her record), she was virtually broke. After a year and a half of full-time employment, she hadn't been able to save one dollar. Other than eighteen months of experience to add to her résumé, Brenda was arguably worse off at the end of the work cycle than she had been at the beginning. She returned to Medicaid and the food stamp line. The promise was a bust.

She became dispirited at first. She didn't follow through when Urban Horizons tried to arrange computer training classes for her. The court dates with Ty's father had a paralyzing effect on her life, more than they probably should have. She felt she couldn't look for a full-time job and deal with family court at the same time. But Brenda was unhampered by a drug addiction, a disability, or chronic depression. When unemployment benefits ran out, Brenda took a job at Saks Fifth Avenue selling elegant hats and pricey scarves to the moderately wealthy during the Christmas season. The job was temporary and ended after New Year's Day.

But then Brenda's darkest cloud finally lifted. On March 5, 2001, after nearly two years of court dates, Brenda won her case against Roosevelt Jenkins. Ty's father brought his pastor to court as a character witness to prove he was a regular churchgoer, a respected Sunday School teacher. He blamed Brenda for his drug use. He claimed he was railroaded into pleading guilty to the rape and beating of the seventy-three-year-old woman years ago in Georgia. He blamed Brenda's daughter for his assault on her when she was fifteen. ("She stabbed me in the leg with a pencil," Teddy told the court. So he punched the teenager in the jaw.) But Bronx Family Court Judge Alma Cordova didn't buy it. She was wary of Teddy's extremely violent past, his admitted drug use, his aggravated harassment with a knife and with his fists. She awarded Brenda one year's worth of protection. A few months later, Brenda won full custody of Ty. "This is it for me. Now I can move on," Brenda said. She had just turned forty-one. Ty was finishing his first-grade year.

More good news followed. Aramark settled her past work grievance, awarding her $2,000 in back pay. By then Brenda had decided she would not allow the pressures of welfare to force her into a go-nowhere job. She'd already made that mistake once. She wanted a job with a future, something with regular raises and benefits, a job that left her alert enough at night to help Ty with his homework. The America Works job search firm no longer operated out of the basement in Brenda's building. WHEDCO had severed most of its ties with the for-profit company.

So Brenda was on her own. She submitted her résumé to the Board of Education for food service jobs. She looked into a union housekeeping position at one of the many downtown hotels. She held out hope for a permanent position at Saks. "I want something I can retire on. Some place that doesn't wear me down." For now, Brenda was still on public assistance. Still scrambling. Still job hunting. Not giving up. She had only a few more months left on her lifetime limit.

Christine's case was the toughest by far. Her family was not only splintered, it had been a damaging force in her fragile life. She was raised by an alcoholic, molested by her mother's friend, turned on to crack by a relative. When her mother died violently, plunging from her apartment window to the ground below, Christine, alone and seventeen, with a baby to care for, was left to carve out her own life from plots of weed and broken glass. She hit bottom at age thirty-five, when she was homeless, jobless, and back on drugs, with no stable family member to boost her. The city removed Christine's children, splitting them up among foster homes. The cycle continued.

Her story of a lifetime of abuse—some of it self-inflicted, some of it not—may be the hardest to understand. But it's an all too familiar story for the majority of welfare recipients as the deadline looms. Obviously, Christine made some inexcusable decisions in her adult life. She knew the homeless shelter guards were watching her every move, yet she took chances, and lost her children in the bargain. She knew her family court date was critical if she were ever to regain her children. Yet she missed the first one and showed up late to the second. Her old addiction had taken hold, once again. What is our responsibility as a society, as a nation, to Christine, or to her children? Are we justified in cutting them off from public help?

Welfare in Christine's life was a necessary drag, something she resorted to when other options ran out. As with Brenda and Alina, the new world of welfare held out rules and hopes that had little to do with Christine's reality. She wanted to work, desperately. But first she had enormous hurdles to overcome before she could imagine organizing her life to sustain a job. Christine had lifelong mental trauma to address, an addiction to overcome, a court case to win, an education to complete. The new welfare rules required more paperwork, more appointments, more work hours than she could master.

It was inevitable, almost cruelly predetermined, that Christine would break one rule or another and be bumped off the rolls, not once but at least three times. The first time welfare cut her off, she was still clean, still working hard to keep herself free of drugs. By the second and third times, she was back to getting high. The strict welfare rules were not to blame for her chaos. Addiction, not welfare, ruled her life. It is logical to argue, however, that the unforgiving regulations added an extra level of static to her already out-of-control life. They may have kept her active, but they did little to address her deep-rooted troubles.

The unraveling accelerated. When she lost her welfare check, she didn't eat. She walked instead of taking the city bus. She became more frantic, not more independent. And she resorted to crime. She gave up. She nearly died.

It's impossible to say whether a less punitive welfare system would have prevented Christine from going around the corner to Cypress Avenue to sell dope to an undercover cop that fateful day in June 1999. Workfare rules and shelter regulations were rarely at the forefront of her mind. Christine lived in a neighborhood teeming with constant temptations. Fighting them off would be difficult enough for a person with a strong family to force her into treatment. She was left to battle her demons virtually alone.

At the nadir of her life, Christine finally found salvation in a drug rehab program, and later in school. Caseworkers at the Samaritan Village removed Christine from the streets of temptation in the Bronx to a safe, wooded lockdown in upstate New York. There she kicked her deadly habit. Welfare paid for her upkeep as long as she was in rehab. The judge monitored her progress. After nine months, Christine moved into a Samaritan Village halfway house in Manhattan's East Fifties. She worked a WEP job at the center's reception desk. Oddly enough, if Christine had entered the system just a year later, these options may not have been available to her. In the final months of the Giuliani administration, the city tightened the rules for recovering addicts on welfare. Addicts now had thirty days to prove they were clean. Relapses could result in benefits being suspended, or cut off completely. Christine's recovery took more than a year, with numerous relapses.

Christine was finally able to visit her girls regularly in their foster home. She reveled in Monica's good grades and her daughter's newfound obsession with a rap music church in Brooklyn. Dyanna, now nine, still ached to come home. She had a much tougher time adjusting to life in limbo. Christine began training to be an ultrasound and diagnostic technician in a technical college on West Sixteenth Street. It's what she had always dreamed of doing. She loved working with patients. It brought out her generous side.

The flush had returned to Christine's face. "I'm really excited about school," Christine said one spring morning, in the waiting area of Manhattan Family Court. Her notebooks were spread out on the seat beside her. She had a test that afternoon in health law. And most encouraging of all, Christine was about to celebrate her one-year anniversary free from drugs. Her once tangled hair was tinted black and coiffed in a neat short wave. She could barely remember the urge to get high that used to rule her waking life. Christine was waiting in court to hear whether and when she could have her daughters returned to her. "I'm gonna get a job in a hospital. I'm gonna get a new apartment, and get my kids back," she said. "Monica inspires me. She tells me she's proud of me."

On March 28, 2001, Judge Joyce Sparrow allowed Christine to see her girls more often and without supervision. "She's doing well," the judge said, casting a smile in Christine's direction. "We need to encourage her all we can." The fleeting kindness caused Christine to choke up for an instant. The Bronx criminal judge was set to consider reducing her drug felony to a misdemeanor in August. Christine had a few more months before welfare kicked her off for good. Government assistance had hung on long enough to keep her alive and bring her family back together.

In the end, when the food stamps were tallied, the rent subsidies totaled, life for Alina, Brenda, and Christine after welfare reform remained very much the same as it was before. Welfare regulations have done little to address domestic violence, poor schools, racial inequities, absent fathers, drug abuse, or low-paying jobs—the complex tapestry of their lives.

• • •

These are the human stories, what skeptics would call anecdotal evidence. Statistical analyses demonstrate a similar mixed bag of numbers five years after welfare reform began.

Besides a severe drop in cases and a surge in employment, the national poverty rate fell for the sixth straight year in 1999 to 12 percent. New York State's rate fell to 15 percent. New York City's poverty rate declined to about 21 percent. Unemployment fell from 7.5 percent nationally in 1992 to 4 percent in 1999. In New York City the unemployment rate declined from 11 percent to 7 percent during the same seven-year period. By the year 2000 it was down to 5.7 percent.

On closer inspection, however, the rosy figures fade. Though the poverty drop was encouraging, it still meant that one in five New Yorkers remained poor. By comparison, European countries such as Germany, France, Great Britain, Norway, Sweden, and Holland registered poverty rates of 5 percent or less. Minorities in New York City were still three times more likely to be poor than whites.

Most would expect to see vast numbers of children emerging from poverty as their parents escaped from the rolls. But that was not the case. Welfare caseloads dropped by 47 percent nationwide by the millennium. Yet the child poverty rate fell only by 17 percent, according to Columbia University's National Center for Children in Poverty. There were one million fewer poor children since 1994. But thirteen million children were still living in poverty, three million more than there had been twenty years ago.

In fact, over the longer term, from 1979 to 1998, the child poverty rates actually increased by over 40 percent in at least a dozen states, including California and New York. Only a handful of states reversed the trend, including New Jersey, Illinois, Arkansas, and South Dakota. Some states reported wildly skewed results, such as Georgia, which cut its welfare load by half and yet saw its child poverty rate soar by 29 percent.

All this was calculated using what most researchers believe is a highly flawed measure—the federal poverty line. The official cutoff poverty point for a family of three was still close to $14,000 at the millennium, no matter where that family lived. Many economic researchers argued that it required more than twice that amount for a family to even think about making ends meet. A recent United Way study found that a working parent in the Bronx with three children would need $38,000 to cover basic expenses, even after tax credits. Another report from the D.C.-based Economic Policy Institute found that it would require more than $40,000 if that same family lived in high-rent urban areas such as Washington, D.C., or San Francisco.

The economic boom of the nineties failed to bridge the chasm between the prosperous and the poor in America. In fact, the gap widened, most starkly in New York City. At the end of the millennium, researchers for the Center on Budget and Policy Priorities determined that wealth was more concentrated among the richest 1 percent of Americans than at any time since the Depression. The top 20 percent of households increased earnings by 40 percent between 1977 and 1999. Income among the bottom fifth of Americans, adjusted for inflation, fell by 9 percent during the same period.

In Brenda's Bronx neighborhood, close to Christine's, the median income was $20,000 in 2001. Thirty-nine percent of Morrisania's population was dependent on welfare, according to the City Project, a local research group. By contrast, on the East Side of Manhattan household income averaged $89,000. Fewer than 1 percent of residents there received welfare.

Other important indicators took an unintended turn for the worse during this period of reform. Some found it more difficult to acquire food and health care. During President Clinton's eight-year tenure, far more people (43 million altogether, including Brenda) went without health insurance. More than half of the 1.2 million former welfare recipients who lost their Medicaid coverage in the late nineties never regained it, according to Families USA, a health care nonprofit research firm.

From 1994 to 1998, 27 percent fewer children (including Alina and her brother) were receiving food stamps nationwide, according to the Food Resource and Action Center, a nonprofit research and public policy group. The following year, Wisconsin turned in the steepest drop in food stamp enrollment of any state in the nation. During roughly the same period, Wisconsin's infant mortality rates rose to 18 percentage points overall. The rate for black infants increased 37 percent. In New York City, emergency food shelters sprouted up to meet the growing demand, from 750 when Mayor Giuliani took office in 1994 to 1,150 when he left eight years later. Homelessness increased at rates unpredicted in welfare reform's early years. Nationwide, 15 percent more families sought shelter, according to the U.S. Conference of Mayors. In New York the numbers were startling. More than twenty-five thousand people required a bed from the city every night in the winter of 2001. The city shelters hadn't seen that population size since the 1980s, when homelessness had reached epidemic proportions.

Child care was another disappointment. Work rules added considerable strain to an already inadequate child care system. A few states used their welfare savings to invest generously in improving care for young children so that single mothers could fulfill their work requirements without worrying about the well-being of their children. But most states did not meet that challenge. A 2000 study by Yale University and University of California at Berkeley researchers found that one million more children were spending their days in child care as a direct result of welfare rules. Most of the care situations ranged from mediocre to seriously disorganized.

In New York City the supply of child care providers never came close to meeting the demand. About one hundred thousand children from infancy to five years of age were not receiving the state-subsidized care for which they were eligible. About nine thousand of those children lived in the Morrisania neighborhood of the Bronx. The Citizens Committee for Children in New York estimated it would take $670 million to make up that difference—money that the public seemed unwilling to spend.

Finally, there was the new phenomenon of the working poor. By 1999 one in six Americans were working and still poor, according to the Urban Institute, a nonpartisan social policy research center. The reason was simple. The average wage for the working poor was $6.16 an hour, not enough by any economic standard to push one adult with two children out of poverty. The National Academy of Sciences reported in the millennium that thirty million Americans lived in families that faced hardships, even if one or more adults were working. In New York City, 43 percent of poor families with children had one worker at home by 1998, according to D.C.'s Fiscal Policy Institute. Over a ten-year period, from 1987 to 1997, the number of working poor grew by 83 percent.

In Wisconsin the numbers were just as stark. A 1999 University of Wisconsin survey found that only one-quarter of those who left the rolls had managed to lift their families above the poverty line. A May 2000 report from the Milwaukee campus of the University of Wisconsin found that the number of children living in working poor families grew 35 percent from 1994 to 1998. One out of three Milwaukee County families with at least one wage earner in the home lived in poverty or near poverty.

National welfare reform dramatically reduced the number of Americans asking for government help. But its promise to improve their lives, and the lives of their children in the bargain, was still an illusion.

B esides requiring work and issuing deadlines, the Personal Responsibility Act allowed states for the first time to spend their welfare money as they saw fit. As a result, most states felt freer to devise their own signature programs. New York installed two unprecedented, industrial-sized programs. One was large-scale fraud detection, the other citywide workfare. Both turned out to be far more effective at bumping recipients off the rolls for violating rules than at transforming public aid recipients into private-sector employees.

Welfare Commissioner Jason Turner recognized these programs' limitations when he took office in 1998. He turned his attention instead to two other, more sweeping reforms. One involved converting the creaky income support centers into "work first" job centers. The other concentrated on hiring private job search companies to match recipients with employment. Both were still designed in part to divert applicants from the welfare lines. But real wage-earning work, even if it was seasonal or minimum-wage, was the goal.

For a time it appeared the commissioner's storied critics would successfully bridle these revolutionary plans. The conversions were stalled in court for years after attorneys for the poor launched a class action suit, charging the centers with sending people in dire need away without emergency food stamps or health care. The second initiative, signing contracts for welfare-to-work companies, found itself tangled up in a federal investigation.

In February 2000, the city comptroller rejected Turner's first contracts with several job search firms. The most widely publicized allegation involved the city's $104 million contract with Maximus, the nation's largest for-profit welfare-to-work company. Turner was familiar with Maximus from his work in the Midwest. The Virginia-based firm had already been cited in a Wisconsin audit for mismanaging state welfare money—pouring funds into advertising and company social events instead of job matches for the poor.

By April 2000, more direct conflicts of interest between the firm and Turner himself began to emerge. Maximus had hired the commissioner's father-in-law to work as a welfare case manager. Turner hired

Tony Kearney, a Maximus employee and a friend of his, as a consultant to HRA at the same time that contract negotiations were under way. Investigators argued that such close access to the commissioner gave the company an unfair advantage over other bidders. In addition, Richard Schwartz, a former Giuliani aide (and chief architect of workfare), was set to receive 30 percent of Maximus's business for his new for-profit Opportunity American job firm.

All this added up to "corruption, favoritism, and cronyism" in the contracting process, according to the city comptroller, Alan Hevesi, also a 2002 city mayoral candidate. Turner adamantly insisted that all the procedures were followed properly. But Hevesi had the last word before turning over his findings to the Manhattan district attorney and federal investigators: "It's as if we were buying a pig in a poke."

A New York Supreme Court judge soon sided with the comptroller, saying there was "compelling evidence the contracting process has been corrupted." A few months later the city's Conflict of Interest Board piled on more charges. The commissioner was fined $6,500 for using HRA employees and equipment to complete work on a private consulting contract. His first deputy, Mark Hoover, was fined $8,500 for leasing apartments he owned to his boss and to various subordinates. The men paid up while their boss, Mayor Giuliani, downplayed the violations, claiming the charges amounted to little more than schoolyard bullying. "I've been in government a long time," said Giuliani, "and this is like Gotcha!"

On October 25, 2000, City Councilman Stephen DiBrienza led dozens of welfare advocates in a chorus of condemnation. Jason Turner should resign immediately, DiBrienza told the crowd, because of his "legacy of corruption and blatant disregard for the law." A litany of pent-up charges spilled forth: Turner's comment that "work will set you free"; his accusation that women make false claims of domestic violence in order to gain benefits; his attempts to turn away the homeless from shelters for missing WEP assignments. The mayor maintained his staunch support. "The more they attack," Giuliani said, "the more I'm probably going to support him."

As it turned out, DiBrienza's censure was ill timed. The same afternoon the appellate court of the State Supreme Court ruled in favor of Turner, saying he did not show favoritism toward Maximus. HRA was simply using an abbreviated contract process, which was justified in this case.

"This is a setback for all the people in this city who have vigorously opposed welfare reform," said Turner's attorney, Michael Hess. "Their position has been beaten down by this decision."

Several months later a federal judge handed Turner another victory. Judge William Pauley III lifted his injunction against the city's plan to convert all its welfare offices into job centers. After two years the judge determined that HRA had become more conscientious about providing emergency Medicaid and food stamps to the desperately sick and hungry. The rest of its violations would be sorted out in a trial to come, but the welfare agency now was following the law better in its new centers than it was in its old ones. The job center conversions lurched forward as planned. It was too soon by the end of Turner's term to determine whether they would be successful.

There were some signs that the city was beginning to recognize the shortcomings of its previous programs. In March 2001, the mayor announced a crash effort to create 517 new rooms for homeless families. Giuliani launched a full-scale low-income housing program in the waning months of his mayoralty.

Yet the reappearance of compassion in his conservative agenda was to be short-lived. Drug abusers came under renewed scrutiny in April 2001. The city's welfare agency announced that recovering addicts had thirty-one days to sober up or risk losing their benefits and city-funded treatment. The same month a State Supreme Court judge found the mayor and his welfare commissioner in contempt of court for failing to help the homeless. Legal aid lawyers brought photos to court showing children and their mothers sleeping on the floor of the Emergency Assistance Unit in the Bronx—the city's last portal to homelessness and welfare endured by Christine and Brenda years before. The practice was illegal. It had been illegal, in fact, for the past fifteen years.

By the time the clock runs out on federal welfare benefits in January 2002, Giuliani will have completed his eight-year term as mayor. Jason Turner was rumored to be considering a position with the new Bush administration in Washington, D.C. Congress will be gearing up to reauthorize new welfare reforms. Any family on welfare when the bill first passed in 1996 will be facing the end of benefits. In New York City, budget analysts estimated that about forty thousand people will still be on the rolls after their lifetime limit runs out.

In some ways these thousands of families are fortunate to live in New York. It was one of only eight states that vowed to pick up some of the welfare costs after the federal government pulled out. But those costs will be greatly reduced. A family of three will receive a grant reduced from $291 to $58.20. A single adult will receive $27 a month. A debit card restricted for use in a few, select stores will be issued for other necessities.

In a surprising hardline stance, the city decided in May 2001 that all those left dangling on the rolls past the five-year lifetime deadline would not automatically receive extensions on their benefits as most state lawmakers intended. Instead, the recipients would have to reapply for welfare, risking delays, at best, or rejection, at worst.

The free ride would be over in December, before a clear destination was ever reached.

In a fit of pique, City Councilman Stephen DiBrienza reacted to this news with a soliloquy on the realities people face on welfare. "Do you suggest they hang around the job office?" he queried Commissioner Turner at a particularly volatile hearing in May. "Do you think they could live with you?" Turner responded by packing up his charts yet again, and leaving the chambers. DiBrienza pelted the commissioner's back with accusations. The two men never found a separate peace.

The success of the 1996 welfare reform act hinged on a set of assumptions about the collective character of the poor. The most prevalent belief was that welfare recipients had lost the work ethic their forebears once held sacred. Government generosity had bred idleness. The poor had become, by this view, a class of people requiring a paternal shove to escape the bonds of public assistance. Stronger character would evolve from old-fashioned labor, regardless of the specific needs and nuances of families' lives.

The philosophy harked back to Victorian beliefs in the redemptive power of work. It was the same notion that tormented my grandfather during the Depression and guided the daily grind of my Protestant family. Those who worked were worthy of government's help, and God's grace. Those who did not work got what they deserved.

Unfortunately, most every policymaker's assumptions on the root causes of poverty have proved misguided in one fashion or another. Nineteenth-century definitions of the "worthy" and "unworthy" poor are no longer useful guidelines. Every worthy beneficiary harbors at least one or twelve unworthy attributes. Every unworthy applicant possesses at least one or twelve positive traits that render her worth saving from herself, or misfortune. Industrial-size programs with industrial-size assumptions about human behavior inevitably overlook the contradictions and textures of people's lives.

Given a choice, nearly everyone would choose meaningful work over a life of idleness. It's a primal instinct, transferred unaltered from Victorian days to twenty-first-century America, from culture to culture, from race to ethnicity. What has changed is the modern economy.

My grandfather left country school after the eighth grade in order to work his first railroad job as a "call boy"—the spry kid who would run through Chaffee, Missouri, before dawn, rapping on windows to wake up the early crew. Like most men born at the turn of the last century, Pat Hancock did not attend high school. Yet after the Depression lifted, he was able to return to his old bookkeeping job at the C&EI Railroad roundhouse. Wages were sufficient to keep his family of eight in new shoes and to furnish the latest gadgets.

In the new millennium, it was nearly impossible for a man or woman with little education to provide as well for such a large family. Subsistence jobs required college degrees, global communication skills, and often Internet expertise. This kind of training was rarely available to those at the bottom end of economic and literacy demographics.

Union protections and guaranteed incomes all but disappeared for the working poor. Instead, welfare reform in the nineties focused almost exclusively on the narrowest of goals—cutting caseloads. The policy presumed that where there was a will, there was a job. Once welfare was eliminated from a person's life, she would find a better way to fill her needs. Deprivation would reignite the American work ethic and thus end poverty. In reality, this policy may have worked for those already experienced and prepared for the twenty-first-century workforce. But for the vast majority, the end of welfare only marked the beginning of a new desperation, and a new form of poverty. Parents now worked long hours for low wages, infants were cared for outside the home, health care was a constant worry, and the family was still poor.

The lesson, ultimately, may be one that politicians are loathe to hear: Public spending on the poor, rather than being the cause of poverty, may provide its greatest hope of a cure. The states with the most successful records in moving the poor into real work have not been the stingiest, but rather the most generous. They have invested in quality child care, preschool, college scholarships, and job training programs. They provided mental health services for welfare recipients. They built low-income housing. Minnesota experimented successfully with subsidizing new workers' low paychecks in order to keep their budgets in the black while their families adjusted. Children's academic records improved. Family health made a comeback in every way.

Before the public grows comfortable with this scope of investment, it probably needs to embrace a new goal for welfare reform—erasing poverty, not just eliminating the dole. In Britain, a voting majority supports the Labor Party's New Deal, which increases welfare spending in order to end child poverty. Single parents are not expected to work. The income gap between the poor and the rich is recognized as a high-priority social problem that affects everyone, regardless of income level. The puzzle of interdependency still baffles American politicians, at a time when our country cannot afford to fracture into further demographic subsets. As Michael Harrington said in his groundbreaking 1962 book, *The Other America: Poverty in the United States*, "If we solve the problem of the other America we will have learned how to solve the problems of all of America."

Afterword

Hands to Work was in production on September 11, 2001, when the world was wrenched apart in a terrifying moment. Everything, everyone, every event, every issue was disfigured in its wake. America's poor, of course, were not immune. The collateral damage was arguably most devastating for those who had the least.

New York City lost nearly 150,000 jobs as a direct result of the World Trade Center attacks. Most jobs were not in high finance. Thousands of hotel workers, vendors, salespeople, messengers, dishwashers, busboys, clerks, and waitresses in lower Manhattan lost positions that provided no cushion, no benefits. Those families that lost undocumented immigrant workers in the tragedy were ineligible to apply for relief. September 11 accelerated the economic recession that had been rumbling under the radar for months.

Beginning in late December, tens of thousands of welfare recipients reached the new federal lifetime welfare limit of five years and were thrust off the rolls into this decimated job market. Newly hired former welfare recipients were the first to be fired, and 70 percent of them were not eligible for unemployment insurance, according to Washington, D.C.'s Economic Policy Institute.

Signs of distress were everywhere. The U.S. Conference of Mayors reported that 23 percent more people were seeking emergency food for Christmas 2001 than had a year earlier. During the same holiday period, 13 percent more people were seeking shelter every night. The generally congenial Twin Cities in Minnesota were turning away 700 people every night in the dead of winter for lack of beds. More than 40 percent of the cities' shelter seekers had jobs—a stark indication that the rise of the working poor was a growing, distressing statistic. New York City officials counted a record 34,000 homeless people looking for a bed every night. More than 40 percent of them were children. The homeless epidemic exceeded the worst of the 1980s.

Several state governors began petitioning Washington to allow them to keep some of their "timed-off" welfare recipients on the rolls. Sending ill-prepared workers into this bleak landscape, they argued, would compound the state's social service woes.

Brenda, Alina, and Christine were caught up in these same roiling waters, struggling to stay afloat. Brenda Fields was herself just months from reaching her five-year lifetime welfare limit on September 11. Her résumé was little changed. She was never able to add the higher education degree or the computer skills she had hoped to attain. Child care was high-priced and hard to find. Scholarships were nonexistent. Brenda had no choice but to join the multitudes of unemployed seeking sales or food service. As Christmas 2001 approached, Brenda said her Baptist prayers amid the Gothic pews at St. Patrick's Cathedral in Manhattan and prayed for a job—any job.

The urgency of her entreaty was palpable. Ty was now a strapping seven-year-old who had grown out of his winter coat and shoes. Her college-aged daughter needed subway fare for commuting to class and work. Brenda was attempting to stretch her family's $180-a-month welfare benefit to cover these needs and her $35-a-month phone bill so that she could set up job interviews. Christmas would be little more than a wish. She had no time to put up decorations no money to buy gifts. Adding to the seasonal stress, her time on welfare was running out. City officials were threatening to cut off her family's benefits if she didn't take a workfare job. Brenda felt as if she had been running in a South Bronx circle for five years.

Eighteen months earlier, Brenda had made the demoralizing trek up the filthy stairs of the Melrose Income Support Center in the Bronx to apply for emergency cash and Medicaid. Aramark had just let her go from her full-time cafeteria job. Welfare was her only cushion. She had no money, no health insurance. Her rent was past due and her blood pressure was in the danger zone. It took about three months for her to find the right medication to get her health in equilibrium. Brenda spent more time figuring out how to pay for medical bills than taking care of her own health. All the while, her custody battle with Tyjahwon's father was creaking through family court. After nearly a year of court dates, the judge finally ruled in Brenda's favor. Brenda was awarded full custody. Ty's father was allowed only professionally supervised visits. After all the angst he had imposed on Brenda, Roosevelt Jenkins arranged no meetings with his son.

After Brenda's unemployment insurance ran its course, she fell back onto full welfare again. This was the last place Brenda imagined being after working full time for nearly two years. She vowed never again to take a low-paying job that would keep her family skating on the edge of poverty. Just as Brenda was well enough to look for a better job, her doctor told her she could no longer afford to put off massive oral surgery. Bone mass in her gums had been slowly disintegrating for years. All her teeth, long neglected as a foster child, were losing their hold. Within a year, they might all be lost. "How could I get a job with no teeth in my mouth?" Brenda wondered. The process of restoring bone and building new teeth would take months of surgery and recovery. Medicaid would cover most of the cost. She would have to rely on friends for the rest.

Meanwhile, the city was pressuring her to sign on for a Work Experience Program (WEP) job, New York's workfare program, or risk losing her benefits. Her own doctor's note was an insufficient excuse. Caseworkers told her she had to see the "welfare doctor," a physician on contract specifically with the city's Human Resources Department to decide whether recipients were physically able to perform workfare labor. The welfare doctor took a look in her scarred and bloodied mouth and gave her a two-month pass from WEP work. Even with the temporary exemption, she was required to visit this doctor twice monthly so he could look in her mouth. For a time, Brenda was seeing the "fraud detector doctor" more often than her own.

By the end of 2001, her surgery was healing. Her patience with the welfare bureaucracy, however, was fraying. She stopped making the mandatory weekly visits to the welfare doctor in Hunts Point. Letters arrived threatening to cut off her benefits. Instead of showing up for her WEP assignment, Brenda applied for a real job in seasonal sales. Saks managers hired her temporarily to sell gloves and scarves for the hoped-for holiday rush, just as they had the previous season. Impressed by her gregarious spirit and inspiring sales record, the store managers this time hired Brenda permanently just before New Year's Day 2002. The timing was fortuitous. She had just been cut off from welfare after failing to show up for WEP duty.

At $12 an hour, Brenda would now be earning just enough to have a few dollars left at the end of the month. She would have to scramble for child care again and sacrifice parenting time with her son. Since the retail industry was hit hard after September 11, her job security was tentative at best. She hoped to make the most of this new position—selling Fendi and Gucci belts to elite Saks customers—so that she could eventually train for something that used her skills as a community organizer.

Despite its myriad caveats, the Saks–belt department job was a happy landing place for the forty-one-year-old mother. Brenda found this position on her own, through the force of her own ambitious personality. If anything unforeseen happened to her family now, she could no longer petition the government for help. Brenda and Ty were now walking the precarious high wire of the shredded job market. Congress had made sure welfare was no longer a safety net for working mothers.

Alina Zukina was a twenty-seven-year-old medical student working her rotation at St. Barnabas Hospital in the Bronx on September 11. Alina's once elusive dream of becoming a doctor was now within reach. Nothing, not even a terrorist attack, could steer her off course. She was one year away from graduating and three more years from finishing her residency in internal medicine. By the time the shy teen from Kishinev turns 30 she will certainly be a licensed physician, poised to begin paying off her $100,000-plus school debt.

Alina thrived on the hard work at St. Barnabas, a rough-edged hospital serving some of the poorest residents of the Bronx. Patients often arrived in advanced stages of an illness or injury. Their symptoms were not ambiguous, yet their conditions were challenging for doctors in training. It was a ripe learning laboratory for the eager student. The hospital was blocks from her new apartment, which she shared with her husband, Mark Filizov—her high school boyfriend from Moldova. After many years of an on-again, off-again engagement, the two Russian refugees finally married. Mark rode the subway every day to midtown, where he worked in pension funds at Alliance Capital, in the Trump Tower. Their lives as ambitious, highly educated immigrants were slowly melting into the American mainstream.

Alina realized she would not be so far along into her education if the city had not helped with food and rent money while she was trying to learn English and study toward a college degree. She survived the maze of welfare regulations to stay on course and stay in college, even when the new rules of welfare conspired against her. Officially, public assistance can no longer be used to support recipients studying for four-year degrees. The city restricts participation in basic education and training programs to only 4 percent of the welfare population. More than half of welfare recipients in New York State have not earned a high school diploma, and 45 percent operate somewhere around a fifth-grade level of literacy.

Alina navigated New York City's restrictive regulations in order to reap the latent value of a higher education degree. Now, in 2002, the dirge-like din of the welfare office and the sweaty encounters with its condescending caseworkers were a buried memory. She had used up her lifetime welfare limit—and then some—by the time she entered medical school on Long Island. It was unlikely the young Jewish immigrant would ever be so desperate to need it again.

Of the three women, Christine Rivera needed to scale the most daunting obstacles to rescue her life and her children's lives from poverty. Her list of catastrophes was numbing. She had been homeless, addicted to heroin, convicted of selling drugs, charged with child neglect. She had lost her children to foster care. At many points in her welfare journey, the Puerto Rican mother was in danger of drowning in the chaos. In an unwitting twist, court-ordered drug rehab may have provided the catalyst that saved her life. Had she thrown her fate in with the welfare rules, she would have been given only a few months of treatment and little tolerance for relapses. Christine needed nearly two years before she could overcome constant relapses to embrace a drug-free life. Her conviction for selling a bag of heroin required her to attend at least eighteen months at a lockdown rehab center in upstate New York. The isolation, and the consistent therapy, provided the tools of resurrection.

Christine moved to a Manhattan halfway house in January 2001. She worked WEP hours at the home's front desk. She took courses in radiology and ultrasound technology in a nearby trade school. After more than a year in foster care, Christine's daughters came home to live with her in the Bronx in June 2001. Christine began internship work toward her medical assistance license at a clinic in the Bronx. Dyanna struggled in a nearby elementary school. Monica thrived as a freshman in Lehman High School. Her adult son, Mark, working as a messenger, married and had a baby boy, Xavier. Christine's youngest child, four-year-old Kristopher, was still living permanently with his father, Luis. On June 28, 2002, Christine received her Registered Medical Assistance License from the State of New York. At $10 an hour, she was now working six days a week at the University Medical Associates, a new clinic affiliated with Bronx Community College. She had been clean for more than two years.

It was still a struggle for Christine to support her large family on this small salary. Federal housing subsidies helped with all but $400 of her rent. Medicaid paid for her children's health care. Her extended family of six was living well below the poverty level. Christine wanted to move everybody from her ramshackle apartment building near Hunts Point. The temptations of the street were too near. There was no time, though, and very little money. With her new certification and her increased experience, Christine hoped to move up in her job and earn more. But she was racing against disaster. Her lifetime welfare limit had all but run out. The power of her desire to do productive work kept her going.

Leaving the welfare rolls was difficult enough, as Brenda and Christine knew. But lifting an entire family out of poverty was entirely another. The two notions were never meant to be mutually exclusive, or so improbable. No question, earning wages was a far more satisfying way of life than living off public assistance. And yet, for the vast majority of former welfare recipients, life off welfare was still a precarious affair. The minimum wage was far too low to support a family, and welfare reform had consistently devalued parenting. According to a four-university study of 700 former welfare mothers and their young children, most families were living in roach-infested housing, skimping on food, and spending less time reading and talking with their children. The April 2002 study, funded in part by the Department of Health and Human Services, noted that poor working mothers suffered from high rates of depression, a condition that had not changed since welfare reform began. Other studies found that adolescents with mothers doing workfare or working at low-paying jobs had more trouble in school than those on welfare.

The welfare bill was scheduled to expire in September 2002. Its reauthorized version needed to provide more financial supplements for low-wage workers, to assure that families were not toiling in poverty, but rather striving for a livable wage. For people like Brenda, it needed to guarantee Medicaid coverage and a real investment in quality child care. For those still on the rolls, like Christine, the bill needed to acknowledge their unique requirements—from mental health care to drug rehabilitation to literacy training and language classes. Job training should be directly attached to real, decent-salaried jobs. Workfare requirements should offer exemptions for mothers with elementary-aged children and for students like Alina who are striving for higher education degrees. Brenda would not have spent five years of her life ducking the go-nowhere workfare jobs if the assignments had been training posts for marketable skills, instead of lessons to recipients on the value of the work ethic.

Perhaps it was too much to hope that September 11 would soften the public's punitive attitude toward the poor and foster a more efficient and compassionate welfare policy for the twenty-first century. For a brief moment, there was promise. In the attack's immediate aftermath, the city set up a joint federal, state, and city relief center on a pier close to ground zero that offered cash and Medicaid for anyone requiring aid. Red tape was instantly untangled. Caseworkers were instructed to help, not divert.

The generosity of post-traumatic New York City lasted only until the shock dissipated. Soon politicians were boasting once again about the great success of the reduced caseloads, ignoring the poor quality of life for the new working poor. By the spring of 2002, President George W. Bush urged Congress to pass even stricter workfare rules. Bush said welfare recipients should work forty hours a week, instead of thirty. States should require 70 percent of their population to perform workfare, up from half. Under the proposal, job training would be slashed to three months from one year. Somehow, single mothers would be expected to fulfill all these requirements without baby sitters. The president added $300 million for untested marriage-boosting programs, but zero dollars for child care. The House of Representatives passed Bush's version, adding $2 billion for child care ($6 billion less than some Democratic senators were lobbying for). As of this writing, it was unclear whether the Senate would weigh in with a slightly more generous version of the welfare bill by the time it expired, or whether it would postpone a vote until 2003.

The prospects for a kinder, gentler policy toward the poor appeared dim. If Bush's plan were approved, more liberal governors feared they would be forced to implement an industrial-sized workfare program similar to New York City's—filled with make-work, menial positions to fulfill federal quotas (fewer than 6 percent of the workers in New York's WEP program ended up with real jobs). Money now spent in more creative job-training programs might be diverted to support the administration of such a giant workfare force. More children might be left with substandard care, because states would not have the funds to create more child care centers. Perhaps worst of all, more women might be forced to marry abusive fathers in order to acquire financial bonuses for their children.

America, the richest nation in the world, could certainly do better for its mothers and children. There is no shame in helping those who have nowhere else to turn.

—LynNell Hancock, July 2002

Reporting Notes

The genesis for this book was sparked by a simple encounter. I was riding a city bus in the Bronx during the winter of 1995, when a boisterous threesome of junior high girls bounded to the seat opposite me, giggling and hollering. As we pulled up to a red light, their happy chatter stopped so abruptly that I looked up from my newspaper. Two friends were prodding the third. "What's up? What's wrong?" they asked their friend. The wiry girl drooped in her seat, her school uniform vest bunched up to her neck, as she looked out the bus window. Outside, a group of five older women dressed in orange work vests were picking up litter around a city garbage can next to the traffic light. "It's Mama," the youngster said. "She told me she was working in an office." Apparently, the mother was too ashamed to tell her daughter she was a WEP worker for the city, picking up trash in the Bronx.

It wasn't until then that I began to take note of these teams of orange-vested workers—single adults, other children's mothers and fathers—cleaning the streets, the parks, City Hall, city courtrooms. I had skimmed over the news stories that touted the steep drop in caseloads, that gave credit to work programs and deadlines. I hadn't realized how detached I was, along with most other middle- and upper-strata New Yorkers, from this startling welfare revolution happening in our streets and in many neighbors' homes.

That bus ride, and the young teen's deep sense of humiliation, made me wonder what human toll lay behind the stream of statistical good news. I thought if I could tell some of the stories about the people behind the vests I

could better understand whether this welfare overhaul was fulfilling its monumental promises.

A couple of years later I began interviewing welfare advocates, lawmakers, politicians, economists, drug counselors, caseworkers, homeless shelter
and settlement house directors, trying to understand the scope of the problem and hoping to find families who might agree to let me invade their lives.
After I decided to narrow my geographic scope to the South Bronx, the hub
of one of the nation's most entrenched welfare populations, my search inten-
sified. I interviewed more than two dozen recipients before Hank Orenstein,
director of a homeless shelter, introduced me to Brenda Fields. She had been
living in the South Bronx shelter with her two children for six months when I
first met her. Brenda's experience with the homeless and welfare systems was
still a raw memory, and she was only too ready to share it with me.

I asked Brenda whether she would be willing to let me stick with her
family for years to come, so I could chronicle her journey through the new
welfare rules. This, obviously, was not a simple request. Scores had turned
me down. Brenda, like so many others I encountered, was in a state of personal crisis. She had lost her job, her home, her sense of stability. Her children's lives were uprooted. Navigating the new rules of welfare often added
stress to her already packed days. I was asking her to include my nosy presence in her life, to record the bare cupboards and cluttered closets of her
experience. I told her I would leave little out. She might have asked what I
could give her in return, but she didn't. Not only did I have nothing tangible
to offer, I brought along with me another set of rules—the rules of journalism. No money. It might taint the reporting process. No direct aid with her
welfare case, or her job search. I was there to record, not to alter, her experience.

I could only offer an abstract vehicle through which to tell her story.
Sometimes the power of the tale is reward enough. That's what Brenda
instinctively understood. Some of her friends warned her against being
exploited by such an arrangement. But Brenda believed in her own story. She
trusted me, for some reason, to help her tell it. She was a generous soul. She
thought welfare policymakers made assumptions about her life and her character that needed some correcting. The details and circumstances of this
unique woman were not unlike the demographics of the typical woman on
welfare—single mother, African American, high school degree, scattered
work history, on and off welfare in the past. So her story could, in a small way,
reflect a larger reality.

I set out to find two more families to round out the textured population.
Russian Jews were entering New York in record numbers, and three-quarters
of them were going on welfare. I was intrigued by the unique circumstances
that immigrants face when they first come to this country. I wondered
whether the new welfare rules influenced their initiation into American life. I

interviewed national organizations that helped Jewish refugees. I visited social service agencies in the Bronx and elsewhere that served Jewish refugees. After several interviews with Russian Jewish welfare recipients, Sue Tozzi, the director of the Bronx Jewish Community Council, introduced me to Alina Zukina, a WEP worker in her office. Alina's story as a Caucasian immigrant student, caught between welfare rules and her lifelong dream to become a doctor, would bring me to another world. She generously agreed to let me into her life, perhaps at first because she didn't realize she could say no.

I entered into these women's stories knowing the beginnings, but not the endings. I had no idea whether Alina would make it into med school, or whether Brenda would find a decent job. It added a dash of anxiety and drama to the reporting journey. I did have a sense, however, that each would find a way to become self-sufficient by sheer force of will. I knew from my reporting that there were thousands of others who probably would not make it off welfare under the deadline wire, the five-year limit. Many had far fewer personal resources than Brenda, and certainly than Alina. These were people suffering from lifelong substance abuse, domestic violence, sexual battery, and mental illness, layered on top of debilitating poverty. Caseworkers told me they were most concerned about this group of people. What would happen to them and their children when their lifetime limit ran out and their wounds were still unhealed? Would our state governments simply refuse to help them, and their children, any longer?

Finding a third family that could help me understand these issues turned out to be my greatest challenge. I visited several day substance abuse treatment centers. I spoke to doctors and practitioners, drug counselors and researchers. I addressed large and small groups of recovering addicts on welfare. I interviewed at least fifty men and women one on one who were still on welfare and trying to stay clean. Their first question was often: "Are you going to pay me?" At first I was shocked to hear that people so readily believed journalists paid subjects to talk to them. Where did that perception come from? Talk shows? Celebrity tabloids? When I tried to fumble through a recitation of the rules of journalism, they would then ask: "Well, will you help me fix my welfare case?" They were searching for any reason on earth to give me their time. I had nothing to offer of any value. The power of journalism to change social conditions was only an anemic afterthought to people with such dire, immediate needs.

I would often return to Hank Orenstein's homeless shelter office at such times, to remember why I was doing this. Then, late into 1998, Hank introduced me to Christine Rivera, a Puerto Rican mother of four who was living in the Jackson Avenue Family Shelter—the same one Brenda had resided in several years earlier. Christine's life was a tangle of chaos, but, Hank said, no more tangled than the lives of most residents he counseled in the shelter.

And Christine filled the room with her bountiful personality. The first day I saw Christine in the shelter, she poured out her meandering story, uncensored, unfettered by everything except her children, who crawled in and out of her lap.

It seemed an easier story for her to tell than it was for me to hear. Christine raced through the outlines of her life as if she were filling out a form. None of it seemed particularly shocking to her. There was craziness in her story, decisions that were clearly outrageous. Jail. Beatings. Too many pregnancies at too young an age. Heroin addiction. A mother killed by accident, or maybe not. An inheritance squandered. A drug-free life derailed. I left the shelter that day overwhelmed by her great flood of catastrophes, some self-inflicted, some beyond her control. Welfare was clearly an asterisk in her life. And maybe that was the point. I wondered how she survived at all, and I wondered how I could translate her chaos onto the page with coherence and honesty. As the interviews progressed, I slowly shed my knee-jerk first impressions. I stopped squirming at some of the foolish decisions in her life and instead tried to understand them in the context of her world. Her story opened my eyes and challenged my craft in ways I never imagined.

My method was the same with all three families. Taking my inspiration from two journalists I greatly admire—Alex Kotlowitz, author of *There Are No Children Here*, and Leon Dash, author of *Rosa Lee*—I traveled back with the women to the beginning of their lives, bringing me chronologically up to date. I interviewed them where they worked and where they lived. As often as I could, I accompanied them to welfare centers, to courtroom hearings, to medical exams, to jail, to classrooms, to their children's schools, so I could get as much firsthand material as possible. In Alina's case, I accompanied her to her homeland of Moldova. I verified their stories with documents, court records, and corroborating interviews. I collected data and studies to help me understand the context of their experiences. Brenda kept every scrap of relevant paper in two large storage bins, which are still in my basement. Alina's paperwork was less voluminous, but better organized. Christine had lost most of her documents, so I relied on the shelter records, official interviews, and detailed incident reports.

This method, immersion reporting, comes with its own set of dilemmas. Intimate reporting over the course of years leads to brutally honest stories, the kinds that explore bright moments as well as dark secrets. Such intimacy often blurs the line between professional journalist and friend. Objectivity is, by nature, dispassionate. Emotional involvement is inevitable, however, while spending days, months, and years with family members, experiencing their jail time, their court appearances, their welfare caseworker frustrations, their graduations, weddings, and birthdays. I strove to keep my professional standards, but not to rein in my empathy. I wanted to understand, not to judge. I often represented the steadiest adult presence in their lives, particularly for

Christine and Brenda, whose own families were dispersed. Perhaps because of this, I found myself interacting with my subjects more than I initially imagined. I wrote a character reference for Brenda's custody case. I took Alina to Moldova. I stored Christine's personal items for her while she was in Rikers. Certainly, I was altering their lives just by being in them.

My most conflicted moments came when I was reporting on Christine. From the time I met her, Christine's life began to unravel. As I found myself witnessing her steady descent into near-death, I wondered whether I could, or even should, intervene. It was frightening, sometimes infuriating, and terribly sad. If I had known then that she would not only kick her addiction, but later thrive in her new school, her new surroundings, it would have been easier. But Christine didn't know what to expect. No one did. Often, when Christine would appear in court, she would scan the room looking for her son, her daughters. Usually, mine was the only familiar face she saw. I could sense both her disappointment and her gratitude. But was my presence doing more harm than good? I tried to tell myself that I might be of some therapeutic use to Christine. With all the tragedy in her life, she always seemed glad to see me. Perhaps just having someone to talk to helped her to some degree. But clearly, the last thing Christine needed was a journalist. She needed a first-rate psychiatrist, a world-class drug program, a steady family, an apartment in a solid community away from the temptations of the South Bronx, decent child care, a college education, a job. She made me doubt my presence, if not my profession of choice.

The most difficult reporting day was when Christine lost her children in Bronx Family Court. I was annoyed with Christine that morning. I couldn't believe she arrived an hour late for this incredibly important court appearance. A judge would decide whether she could keep her children or not, and her hair was wet from a last-minute shower. I didn't understand then that she was already deep into the grip of heroin, a week after her children had been removed to foster care. She had been picked up for shoplifting a few days earlier to pay for her habit. A caseworker came up to me at the end of the long ordeal in court that day with a list of orders. I should stay with Christine all day, she said, in order to make sure she didn't do drugs. I should take her to a certain residential drug treatment center to enroll her in the program immediately, whether she wanted to or not. Christine was vulnerable. She had no one else. I felt helpless. I was not supposed to be Christine's tough-love parent. "I'm just a journalist," I remember hearing myself tell the caseworker, choking on my own words. "I can't force her to do anything against her will." Christine did not want to come to lunch with me. She didn't want to go to the treatment center. She just wanted to go into her room in the shelter and shut the door. So I left, wondering again whether I should switch careers.

The ethical dilemmas were not all steeped in such tragedy. Shortly after Brenda was hired at Aramark food corporation full-time, she surprised me with a touching gift. It was classic Brenda. The moment she had a little money to spend, she wanted to spread her good feelings all around. Inside the bag was a smart wool blazer just my size that she had picked up at her favorite thrift shop. I spent an awkward moment debating in my head whether I should accept it. The rules of professional conduct dictated no gifts, no money exchanged between journalist and subject. I had bought birthday and Christmas gifts for Ty before, but not for his mom. It would have crushed Brenda, however, if I rejected the blazer. I also knew the transaction would not compromise her story, already written. After two years I no longer had any doubt about the integrity of my reporting. I accepted this gift from a valued friend, and we have exchanged others for birthdays and holidays ever since.

The bulk of my book was finished when Brenda lost her Aramark job. She had no savings. She was sick. There was no money for food or back rent. She had applied for welfare and for unemployment benefits, but neither would kick in for at least a month. Brenda was always the first person to help out any neighbor with money or baby-sitting. Now she was desperate. She didn't ask me for help. She never had. But I gave her enough money to cover food and rent and medicine until her benefits arrived. Perhaps I broke a journalist's code with that gift. But Brenda's health and her children's well-being hung in the balance. I looked at her son Ty and recognized in his plight a bit of the junior high girls on the Bronx bus all over again. This time I knew the child too well. I couldn't just observe.

Selected Sources

Abramovitz, Mimi. *Under Attack, Fighting Back: Women and Welfare in the United States*. New York: Monthly Review Press, 1996.

———. *Regulating the Lives of Women: Social Welfare Policy from Colonial Times to the Present*. Boston: South End Press, 1996.

Accles, Liz and Liz Krueger. *Workfare: The Real Deal*. New York: Community Food Resource Center, June 18, 1996.

Acs, Gregory, Katherine Ross Phillips, and Daniel McKenzie. *Playing by the Rules but Losing the Game: America's Working Poor*. Washington, D.C.: Urban Institute, May 2000.

Addams, Jane. *Twenty Years at Hull-House*. Boston: Bedford/St. Martin's, 1999.

Allen, Michael O., and Dave Saltonstall. "Workfare Not a Killer—Rudy." *New York Daily News*, February 23, 1997.

America's Homeless Children: New Outcasts. Newton, Mass.: Better Homes Fund, 1999.

American Community Survey Profile for Bronx County, New York, 1999. Washington, D.C.: U.S. Bureau of the Census, 1999.

Arnett, Elsa C. "Little Help for Recipients as Some States Wrangle." *San Jose Mercury News,* February 29, 2000.

Bai, Matt, and Gregory Beals. "A Mayor Under Siege." *Newsweek,* April 5, 1999.

Bane, Mary Jo, and David T. Ellwood. *Welfare Realities: From Rhetoric to Reform*. Cambridge, Mass.: Harvard University Press, 1994.

Barr, Heather. *Prisons and Jails: Hospitals of Last Resort*. New York: Correctional Association of New York/Urban Justice Center, 1999.

Barrett, Wayne. "Fifty Reasons to Loathe Your Mayor." *Village Voice*, November 4, 1997.

———. *Rudy: An Investigative Biography of Rudolph Giuliani*. New York: Basic Books, 2000.

Bearak, Barry. "A Mercurial Mayor's Confident Journey." *New York Times*, October 19, 1997.

Bernstein, Nina. "Poverty Rate Persists in City Despite Boom." *New York Times*. October 7, 1999.

———. "Work-for-Shelter Requirement Is Delayed by New York Judges." *New York Times*, December 9, 1999.

———. "Giuliani Proclaims Success on Pledge to Curb Welfare." *New York Times*, December 29, 1999.

———. "Widest Income Gap Is Found in New York." *New York Times*, January 19, 2000.

———. "Welfare Plan in City Suffers a Setback." *New York Times*, February 3, 2000.

———. "Family Needs Far Exceed the Official Poverty Line." *New York Times*, September 13, 2000.

———. "Homeless Shelters in New York Fill to Highest Level Since '80s." *New York Times*, February 8, 2001.

Berrick, Jill Duerr. *Faces of Poverty*. New York: Oxford University Press, 1995.

Bohlen, Celestine. "How Mother Russia Plucks Her Pensioners Clean." *New York Times*, March 17, 1999.

Bos, Hans, Aletha Huston, Robert Granger, Greg Duncan, Tom Brock, and Vonnie McLoyd. *New Hope for People with Low Incomes: Two-Year Results of Program to Reduce Poverty and Reform Welfare*. New York: Manpower Demonstration Research Corporation, April 1999.

Bradley, James. "The Turner Diaries." *Village Voice*, January 27, 1998.

Bumiller, Elisabeth. "For Giuliani, Making Most of Artful War with Elite." *New York Times*, October 16, 1999.

———. "Shelters Vow to Defy Mayor on Work Rule." *New York Times*, December 18, 1999.

Burrows, Edwin G., and Mike Wallace. *Gotham: A History of New York City to 1898*. History of New York City Series. New York: Oxford University Press, 2000.

Bush, Andrew. "Replacing Welfare in Wisconsin." Briefing Paper. Indianapolis: Hudson Institute, July 1996.

Bush, Andrew, Swati Desai and Lawrence Mead. "Leaving Welfare: Findings from a Survey of Former New York City Welfare Recipients." Working Paper. New York: Human Resources Administration, September 1998.

Caro, Robert A. *The Power Broker: Robert Moses and the Fall of New York*. New York: Vintage Books, 1975.

Casey, Timothy J. *Welfare Reform and Its Impact in the Nation and in New York*. New York: Federation of Protestant Welfare Agencies, August 1998.

———. *NY Welfareform Digest 32: Poverty*. New York: Federation of Protestant Welfare Agencies, February 7, 2001.

Center for the Future of Children. *The Future of Children: Welfare to Work. David and Lucile Packard Foundation,* 7, no. 1 (Spring 1997).

Chan, Ying, and K. C. Baker. "Workfare Ed Rules Blasted." *New York Daily News,* February 25, 1997.

Child Care: The Family Life Issue in New York City. New York: Citizens' Committee for Children, May 2000.

"Child Poverty in the States: Levels and Trends from 1979 to 1998." Childhood Poverty Research Brief 2. New York: National Center for Children in Poverty, Mailman School of Public Health, Columbia University, 2000.

Clinton, Bill, and Al Gore. *Putting People First: How We Can All Change America.* New York: Times Books, 1992.

Dash, Leon. *Rosa Lee: A Mother and Her Family in Urban America.* New York: Basic Books, 1996.

DeMause, Neil. "Move 'em Out." *In These Times,* November 3, 1997–December 14, 1997.

DeParle, Jason. "Success, and Frustration, as Welfare Rules Change." *New York Times,* December 30, 1997.

———. "Getting Opal Caples to Work." *New York Times Magazine,* August 24, 1997.

———. "Faith in a Moral Motive for Work; The Man Who Redesigned Welfare in Wisconsin Is Coming." *New York Times,* January 20, 1998.

———. "What Welfare-to-Work Really Means." *New York Times Magazine,* December 20, 1998.

———. "Spending the Savings; Leftover Money for Welfare Baffles, or Inspires, States." *New York Times,* August 29, 1999.

———. "As Benefits Expire, the Experts Worry." *New York Times,* October 10, 1999.

———. "Early Sex Abuse Hinders Many Women on Welfare." *New York Times,* November 28, 1999.

———. "A War on Poverty Subtly Linked to Race." *New York Times,* December 26, 2000.

———. "Bold Effort Leaves Much Unchanged for the Poor." *New York Times,* December 30, 1999.

Dolkart, Andrew S. *Guide to New York City Landmarks: New York City Landmarks Preservation Commission.* New York: John Wiley & Sons, 1998.

Downside: The Human Consequences of the Giuliani Administration's Welfare Caseload Cuts. New York: Federation of Protestant Welfare Agencies, November 2000.

Drew, Christopher, and Eric Lipton. "Two With Ties to Chief of Welfare Got Jobs with Major Contractor." *New York Times,* April 21, 2000.

Duncan, Greg J., and Jeanne Brooks-Gunn, eds. *Consequences of Growing up Poor.* New York: Russell Sage Foundation, 1997.

Edelman, Peter. *Searching for America's Heart: RFK and the Renewal of Hope.* Boston: Houghton Mifflin, 2001.

Edin, Kathryn, and Laura Lein. *Making Ends Meet: How Single Mothers Survive Welfare and Low-Wage Work.* New York: Russell Sage Foundation, 1997.

Fernandez, Humberto. *Heroin.* Center City, Minn.: Hazelden, 1998.

Fifield, Adam. "Stop Payment: Cuts in Disability Benefits and Food Stamps Threaten the Survival of Legal Immigrants." *City Limits,* April, 1997.

Finder, Alan. "Evidence Is Scant That Workfare Leads to Full-time Jobs." *New York Times,* April 12, 1998.

Firestone, David. "Without Welfare, Giuliani Suggests, Many Might Move." *New York Times,* April 29, 1995.

———. "Clinton to Sign Welfare Bill That Ends U.S. Aid Guarantee and Gives States Broad Power." *New York Times,* August 1, 1996.

———. "Debating Workfare." *New York Times,* August 21, 1996.

———. "Mayor Defends Workfare for Students." *New York Times,* March 14, 1997.

———. "Praising the Wonders of Workfare, Giuliani Finds a Campaign Theme." *New York Times,* March 20, 1997.

Fitzpatrick, Joseph P., S.J. *The Stranger Is Our Own: Reflections on the Journey of Puerto Rican Migrants.* Kansas City, Mo.: Sheed and Ward, 1996.

Freedberg, Louis. "The Refugee Sweepstakes: Who Gets in, Who's Left Out." *San Francisco Chronicle,* August 19, 1994.

Fuentes, Annette. "Slaves of New York." *In These Times,* December 23, 1996.

Fuller, Bruce, and Sharon Lynn Kagan. *Remember the Children: Mothers Balance Work and Child Care Under Welfare Reform.* Growing up in Poverty Project, University of California at Berkeley: February 2000.

Funiciello, Theresa. *Tyranny of Kindness: Dismantling the Welfare System to End Poverty in America.* New York: Atlantic Monthly Press, 1993.

Gans, Herbert J. *The War Against the Poor: The Underclass and Antipoverty Policy.* New York: Basic Books, 1995.

Gilens, Martin. *Why Americans Hate Welfare.* Chicago: University of Chicago Press, 1999.

Giuliani, Rudolph W. "Immigration: The Progress We've Made and the Road Ahead." Speech delivered in New York City, March 31, 1998.

———. "Reaching out to All New Yorkers by Restoring Work to the Center of City Life." Speech delivered in New York City, July 20, 1998.

———. "The Next Phase of Quality of Life: Creating a More Civil City." Speech delivered in New York City, February 24, 1998.

———. "An Agenda to Prepare for the Next Century." Speech delivered in New York City, January 14, 1999.

———. "Reinforcing the Social Contract and the Work Ethic." Speech delivered in New York City, November 7, 1999.

Gorin, Julia. "Russians Get off Rolls, with Help." *New York Daily News,* April 13, 1997.

Green, Mark. *New York City "Way"—Not Yet the Way: A Preliminary Examination of the New NYC Program to Reduce Home Relief Fraud and Encourage Independence.* New York: Office of the Public Advocate, June 1995.

Greene, Liza M. *New York for New Yorkers: A Historical Treasury and Guide to the Buildings and Monuments of Manhattan.* New York: W. W. Norton, 1995.

Greenfield, Lawrence A., and Tracy L. Snell. *Women Offenders.* Bureau of Justice Statistics Special Report. Washington, D.C.: U.S. Department of Justice, December 1999.

Gullo, Karen. "States Hoarding $7 Billion in Unspent Welfare Money." Associated Press, February 24, 2000.

Handler, Joel F., and Yeheskel Hasenfeld. *We the Poor People: Work, Poverty, and Welfare*. New Haven, Connecticut: Yale University Press, 1997.

Harden, Blaine. "Giuliani Seeks an End to Welfare; New York City Mayor Wants to Halt Aid Completely by 2000." *Washington Post*, July 21, 1998.

Harrington, Michael. *The Other America: Poverty in the United States*. New York: Touchstone Press, 1992.

Hernandez, Donald J., ed. *Children of Immigrants: Health, Adjustment, and Public Assistance*. Washington, D.C.: National Academy Press, 1999.

Hernandez, Raymond. "Most Dropped from Welfare Don't Get Jobs." *New York Times*, March 23, 1998.

———. "Most Off Welfare Get Jobs, But Not All Long-term." *New York Times*, August 5, 1999.

———. "Court Rejects Giuliani's Policy on AIDS Benefits." *New York Times*, October 20, 1999.

———. "Albany Funds for Child Care Going Unspent." *New York Times*, October 25, 1999.

———. "New York Gets Big Windfall from Welfare." *New York Times*, February 9, 1999.

———. "U.S. Welfare Limit May Put Thousands in Albany's Care." *New York Times*, March 21, 2000.

Herr, Toby, Suzanne Wagner, and Robert Halpern. *Making the Shoe Fit: Creating a Work-Prep System for a Large and Diverse Welfare Population*. Chicago: Project Match/Erickson Institute, December 1996.

Hicks, Jonathan. " 'White Guys' Memo to Mayor Arouses Anger." *New York Times*, February 13, 1994.

"Homelessness: Programs and the People They Serve." *National Survey of Homeless Assistance Providers and Clients*, December 7, 1999.

Horn, Wade, and Andrew Bush. "Fathers, Marriage, and Welfare Reform." Indianapolis: Welfare Policy Center, Hudson Institute, March 1997.

Hunger Is No Accident: New York and Federal Welfare Policies Violate the Human Right to Food. New York: Welfare Reform and Human Rights Documentation Project, July 2000.

Jackson, Kenneth T. *Encyclopedia of New York City: New-York Historical Society*. New Haven, Conn.: Yale University Press, 1995.

Johnson, Nicholas. *A Hand Up: How State Earned Income Tax Credits Help Working Families Escape Poverty*. Washington, D.C.: Center on Budget and Policy Priorities, 1999.

Johnston, David Cay. "Gap Between Rich and Poor Found Substantially Wider." *New York Times*, September 5, 1999.

Judge, Edward H. *Easter in Kishinev: Anatomy of a Pogrom*. New York: New York University Press, 1992.

Kahn, Alfred J., and Sheila B. Kamerman. *Big Cities in the Welfare Transition*. New York: Cross-National Studies Research Program, Columbia University School of Social Work, 1998.

Kaplan, Robert D. *Balkan Ghosts: A Journey Through History*. New York: St. Martin's Press, 1993.

Katz, Michael B. *In the Shadow of the Poorhouse: A Social History of Welfare in America*. New York: Basic Books, 1986.

———. *The Undeserving Poor: From the War on Poverty to the War on Welfare*. New York: Pantheon Books, 1989.

Kearns Goodwin, Doris. *No Ordinary Time: Franklin and Eleanor Roosevelt: The Home Front in World War II*. New York: Simon and Schuster, 1995.

Keeping Track of New York's Children. New York: Citizens' Committee for Children, 1993, 1995, 1997, 1999.

Kessner, Thomas. *Fiorello H. La Guardia and the Making of Modern New York*. New York: McGraw-Hill, 1989.

Kids Count Data Book. Baltimore: Annie E. Casey Foundation, 1996, 1997, 1998, 1999, 2000.

Kirschenbaum, Jill. "The Long Road Home." *City Limits*, November 1993.

Kotlowitz, Alex. *There Are No Children Here: The Story of Two Boys Growing up in the Other America*. New York: Anchor Books, 1991.

Kozol, Jonathan. *Amazing Grace: The Lives of Children and the Conscience of a Nation*. New York: HarperPerennial, 1995.

———. *Ordinary Resurrections: Children in the Years of Hope*. New York: HarperPerennial, 2001.

Krueger, Liz, Liz Accles and Laura Wernick. *Workfare: The Real Deal II*. New York: Community Food Resource Center, July 1997.

Lakhman, Marina, and David Goldiner. "Beacon of Liberty for New Russians Filled with Ambition, Immigrants Find Success in the City." *New York Daily News,* January 31, 1999.

Legacy of Neglect: The Impact of Welfare Reform on New York's Homeless. New York: Coalition for the Homeless, August 1999.

Lemann, Nicholas. *The Promised Land: The Great Black Migration and How It Changed America*. New York: Vintage Books, 1992.

Lerman, Robert I. *Retreat or Reform? New U.S. Strategies for Dealing with Poverty*. Washington, D.C.: Urban Institute, 1999.

Lerman, Robert I., Pamela Loprest, and Caroline Ratcliffe. *How Well Can Urban Labor Markets Absorb Welfare Recipients?* New Federalism Issues and Options for States, series A, no. A-33. Washington, D.C.: Urban Institute, June 1999.

Levy, Clifford J. "Immigrant's Death Becomes an Issue in Mayoral Race." *New York Times,* February 23, 1997.

Liff, Bob, and Russ Buettner. "City Expanding Workfare for Single Parents." *New York Daily News,* March 19, 1996.

Lipton, Eric. "Social Services Official Admits He Broke Conflict-of-Interest Law and Is Fined." *New York Times,* October 17, 2000.

———. "Rejecting Favoritism Claim, Court Upholds a City Welfare Contract." *New York Times,* October 25, 2000.

———. "City Revises Plan to Create Jobs for Poor." *New York Times,* November 21, 2000.

————. "In Settlement, Welfare Office Will Become City Job Centers." *New York Times,* February 15, 2001.

————. "Judge Orders City to Shelter Two Suffering Homeless Families." *New York Times,* April 12, 2001.

Loprest, Pamela. *How Families That Left Welfare Are Doing: A National Picture.* New Federalism National Survey of America's Families. Washington, D.C.: Urban Institute, August 1999.

Losing Health Insurance: The Unintended Consequences of Welfare Reform. Washington, D.C.: Families USA Foundation, May 1999.

Lueck, Thomas J. "Hours Before Court Deadline, City Agrees to Add Rooms for Homeless." *New York Times,* March 1, 2001.

————. "Hevesi Rejects Largest Pacts for Workfare." *New York Times,* March 23, 2000.

————. "Workfare Critics Cite the Case of a Woman Who Died on the Job." *New York Times,* August 31, 2000.

Mann, Arthur. *La Guardia Comes to Power: 1933.* Westport, Conn.: Greenwood Publishing Group, 1981.

Mauer, Marc, Cathy Polter, and Richard Wolf. *Gender and Justice: Women, Drugs, and Sentencing Policy.* Washington, D.C.: Sentencing Project, November 1999.

Mayor's Management Report. New York: City of New York, fiscal 1996, 1997, 1998, 1999, 2001.

McCormick, John, and Evan Thomas. "One Family's Journey from Welfare to Work." *Newsweek,* May 26, 1997.

McCoy, Kevin. "From Old World to New Millennium: As Families Reunite in a Land of Promise, the City's Face Changes." *New York Daily News,* November 26, 1999.

McMillan, Tracie. "City Tells Drug Users on Welfare: Sober up or Else." *City Limits,* April 9, 2001.

Mead, Lawrence M. *The New Politics of Poverty: The Nonworking Poor in America.* New York: Basic Books, 1992.

————. "Conflicting Worlds of Welfare Reform." *First Things* 75 (August–September 1997): 15.

————. ed. *The New Paternalism: Supervisory Approaches to Poverty.* Washington, D.C.: Brookings Institution Press, 1997.

"Medicaid: Early Implications of Welfare Reform for Beneficiaries and States." Washington, D.C.: U.S. General Accounting Office, February 1998.

"Medicaid Enrollment: Amid Declines, State Efforts to Ensure Coverage After Welfare Reform Vary." Washington, D.C.: U.S. General Accounting Office, September 1999.

Mink, Gwendolyn. *Welfare's End.* Ithaca, N.Y.: Cornell University Press, 1998.

"Moldovan Power Provider Negotiates New Electric Supply from Romania." *Welcome,* May 27, 1999.

Nathan, Richard P., and Thomas L. Gais. *Implementing the Personal Responsibility Act of 1996: A First Look.* Albany: Nelson A. Rockefeller Institute of Government, Federalism Research Group, State University of New York, 1999.

The Newest New Yorkers, 1995–1996: An Analysis of Immigration to New York City in the Early 1990s. New York: Department of City Planning, December 1996.

Nolan, Clare. "Minnesota's Welfare Program Shows Dramatic Results." www.stateline.org, May 31, 2000.

"Obstructing Local Welfare Reform" (editorial). *New York Times*, July 4, 1998.

O'Neill, Hugh, Kathryn Garcia, and Kathryn McCormick. *Where the Jobs Are: How Labor Market Conditions in the New York Area Will Affect the Employment Prospects of Public Assistance Recipients*. New York: Community Service Society, April 1997.

Opportunities for Change: Lessons Learned from Families Who Leave Welfare. New York: Citizens' Committee for Children, January 2000.

Parrott, Sharon. *Welfare Recipients Who Find Jobs: What Do We Know About Their Employment and Earnings?* Washington, D.C.: Center on Budget and Policy Priorities, November 1998.

The Pataki Welfare Plan. New York: Association of the Bar of the City of New York, December 1996.

Pear, Robert. "Clinton Plan to Seek out Those Eligible for Food Stamps." *New York Times*, July 14, 1999.

———. "Gains Reported for Children of Welfare-to-Work Families." *New York Times*, January 23, 2001.

Piven, Frances Fox, and Richard A. Cloward. *The Breaking of the American Social Compact*. New York: New Press, 1967.

———. *Regulating the Poor: The Functions of Public Welfare*. New York: Vintage, 1971.

Pollitt, Katha. "Let Them Sell Lemonade." *The Nation*, February 15, 1999.

Polner, Robert. "Rudy Hires a Welfare Reformer." *Newsday*, January 8, 1998.

———. "Rudy Weighs New Welfare Rules for Single Parents." *Newsday*, May 12, 1998.

———. "Work and Welfare: New Rules Aim to Reduce Number on City's Rolls." *Newsday*, November 1, 1998.

———. "Workfare Program Feels the Heat: Worker's Death Prompts Outcry." *Newsday*, July 31, 1999.

Polner, Robert, and Michael Powell. "It Isn't Working; Many Want Jobs That Aren't There." *Newsday*, March 30, 1995.

Poverty, Welfare, and Children: A Summary of the Data. Washington, D.C.: Child Trends Research Brief, 1999.

Pressman, Gabe. Interview with Rudolph Giuliani and Ronald Lauder. *News 4orum*, WNBC-TV News 4, New York, September 3, 1989.

Ramirez, Anthony. "Rise in New York City Children Who Are Born into Poverty." *New York Times*, February 4, 1999.

Rank, Mark Robert. *Living on the Edge: The Realities of Welfare in America*. New York: Columbia University Press, 1994.

Rationing Charity: New York City Struggles to Keep up with Rising Hunger: The 1998 Survey of New York City Emergency Food Programs. New York: New York City Coalition Against Hunger, October 15, 1998.

Redburn, Tom. "Conservative Thinkers Are Insiders; It's Now Their City Hall, and Manhattan Institute Is Uneasy." *New York Times*, December 31, 1993.

Reeves, Phil. "Russian Jews Fear Return of Pogroms." *The Independent* (London), March 7, 1998.

Republic of Moldova. Chisinau: Litera Moldpres, 1998.

Riedinger, Susan A., Laudan Aron, Pamela Loprest, and Carolyn O'Brien. *Highlights from State Reports: Income Support and Social Services for Low-Income People in New York*. Washington, D.C.: Urban Institute, 1998.

———. *Income Support and Social Services for Low-Income People in New York*. Washington, D.C.: Urban Institute, 1998.

Sánchez Korrol, Virginia E. *From Colonia to Community; The History of Puerto Ricans in New York City*. Berkeley: University of California Press, 1994.

Schleuning, Neala. *Idle Hands and Empty Hearts: Work and Freedom in the United States*. New York: Bergin and Garvey, 1990.

Schor, Juliet B. *The Overworked American: The Unexpected Decline of Leisure*. New York: Basic Books, 1992.

Schwartzman, Paul. "Driven from the Start." *New York Daily News*, May 11, 1997.

———. "Politics over Priesthood." *New York Daily News*, May 12, 1997.

———. "Love and the Law." *New York Daily News*, May 13, 1997.

Seefeldt, Kristin S., Laura Kaye, Christopher Botsko, Pamela A. Holcomb, Kimura Flores, Carla Herbig, and Karen C. Tumlin. *Highlights from State Reports: Income Support and Social Services for Low-Income People in Wisconsin*, 1998.

Sengupta, Somini. "Hundreds Lose Food Stamps in Error Under U.S. Welfare Change." *New York Times*, March 9, 1998.

———. "State's Poorest Facing Loss of U.S. Aid." *New York Times*, February 10, 2001.

Shore, Rima. *Our Basic Dream: Keeping Faith with America's Working Families and Their Children*. New York: Foundation for Child Development, October 2000.

Siegel, Fred. *The Future Once Happened Here: New York, D.C., L.A., and the Fate of America's Big Cities*. New York: Free Press, 1997.

Snapshots of America's Families: A View of the Nation and Thirteen States. Washington, D.C.: Urban Institute, January 1999.

Snapshots of America's Families II: A View of the Nation and Thirteen States. Washington, D.C.: Urban Institute, October 2000.

Solow, Robert M., and Amy Gutmann, eds. *Work and Welfare*. Princeton, N.J.: Princeton University Press, 1998.

The State of America's Children Yearbook 2000. Washington, D.C.: The Children's Defense Fund, 2000.

The State of Municipal Services in the 1990s: Social Services in New York City. New York: Citizens' Budget Commission, August 1997.

A Status Report on Hunger and Homelessness in American Cities: A Thirty-City Survey. Washington, D.C.: U.S. Conference of Mayors, December 1998.

Steps to Success: Helping Women with Alcohol and Drug Problems Move from Welfare to Work. New York: Legal Action Center, May 1999.

Stettner, Andrew. *Welfare to Work: Is It Working?* Washington, D.C.: Georgetown University Graduate Policy Institute/Community Voices Heard, 1999.

"Study Reveals Working Families Are Locked out of Child Care." *Children's Defense Fund Reports* 19, nos. 4 and 5 (April–May 1998).

Swarns, Rachel. "The 1997 Elections: The Issues; 320,000 Have Left Welfare, but Where Do They Go from Here?" *New York Times,* October 29, 1997.

———. "Wisconsin Welfare Chief Chosen for New York City." *New York Times,* January 8, 1998.

———. "Giuliani to Place Disabled Mothers in Workfare Jobs." *New York Times,* June 8, 1998.

———. "U.S. Inquiry Asks If City Deprives Poor." *New York Times,* November 8, 1998.

———. "U.S. Audit Is Said to Criticize Giuliani's Strict Welfare Plan." *New York Times,* January 20, 1999.

———. "Judge Delays Giuliani Plan on Welfare." *New York Times,* January 26, 1999.

Task Force for Sensible Welfare Reform. New York: Robert J. Milano Graduate School of Management and Urban Policy, New School for Social Research, May 1996.

———. *Welfare Reform in New York: A Report on Implementation Issues in New York City. Task Force for Sensible Welfare Reform.* Robert J. Milano Graduate School of Management and Urban Policy, New School University, January 1999.

Tierney, John. "The Greatest? Give Mayor a Mirror." *New York Times,* January 15, 2000.

Turner, Jason. "Work Can Restore Lives." *New York Daily News,* March 7, 1999.

Ultan, Lloyd. *The Beautiful Bronx: 1920 to 1950.* New York: Harmony Books, 1979.

The Unfinished Business of Welfare Reform: Fixing Government Policies That Exclude Working Poor from Benefits. New York: Community Service Society, November 1999.

Vergara, Camilo Jose. *The New American Ghetto.* New Brunswick, N.J.: Rutgers University Press, 1997.

Vital, David. *A People Apart: The Jews in Europe, 1789–1939.* Oxford: Oxford University Press, 1999.

Vogel, Carl, and Neil deMause. "Jason's Brain Trust." *City Limits,* December 1998.

Wagenheim, Kal, and Olga Jimenez de Wagenheim, eds. *The Puerto Ricans: A Documentary History.* Princeton, N.J.: Markus Wiener Publishers, 1994.

Weiner, Miriam. *Jewish Roots in Ukraine and Moldova: Pages from the Past and Archival Inventories.* Secaucus, N.J. and New York: Miriam Weiner Routes to Roots Foundation and YIVO Institute for Jewish Research, 1999.

Wertheimer, Richard F. *Working Poor Families with Children.* Washington, D.C.: Child Trends, February 1999.

Wilson, William Julius. *When Work Disappears: The World of the New Urban Poor.* New York: Knopf, 1996.

Young Children in Poverty: A Statistical Update, March 1998. New York: National Center for Children in Poverty, Mailman School of Public Health, Columbia University, 1998.

Young Children in Poverty: A Statistical Update, June 1999. New York: National Center for Children in Poverty, Mailman School of Public Health, Columbia University, 1999.

Acknowledgments

Many people helped me piece this book together, both directly and indirectly. My deepest appreciation goes to the three women who allowed me to insinuate myself into their lives. The book could not have been written without their generosity and unflagging honesty. Between study and work, Alina Zukina barely had time to sigh, yet she stayed with this project through long months and stressful times. I am grateful to her parents, her brother, and her fiancé, Mark, for letting Alina accompany me to Moldova.

Christine always found the stamina to share her sharp intelligence with me during one of the darkest periods of her life. It is particularly gratifying to know that Christine made such a remarkable recovery, most notably after the book was finished. Brenda provided the moral sustenance for the book. She is a fierce optimist, an inspired mother, and one of the most charitable people I know. The indomitable Tyjahwon and his elegant sister are testaments to her persistence. It took courage for all these women and their families to let a stranger tell their stories. I hope I've done them justice.

Spouses are usually shuttled to the last paragraph in acknowledgments, but my quirky sportswriter soul mate and sparring partner deserves top billing. Little did Filip Bondy know what he was getting into when he encouraged me to become a journalist a hundred years ago. Nobody is as smooth and as cranky a word doctor as the Bleacher Creature. I owe him pretty much everything, including my half of the mortgage.

New York University student Tracie McMillan spent more months than I can count chasing down reports and statistics and legislation, and then chasing down people who could tell her what the reports and statistics and legislation mean. Tracie's research is woven into nearly every page of the book.

Hundreds of people in the South Bronx and beyond helped shape this book by brainstorming with me or submitting to formal interviews. I mention only a few: Hank Orenstein, Joseph Esheyigba, and Wanda Abeyllez at the Jackson Avenue Family Shelter, who introduced me to Brenda and Christine and the gateway to welfare posed by the homeless system. At Urban Horizons, codirectors Nancy Biberman and Barbara Petro-Budacovich provided inspiring examples and important perspectives on the lives of recipients. Liz Krueger and Don Friedman of the Community Food Resources Center offered clearheaded insights and hard research to help me understand the vicissitudes of New York City's reforms.

I'm indebted to many academics, researchers, and journalists whose work helped shape my thinking. They include Barbara Blum and Larry Aber of the National Center for Children in Poverty, Jeanne Brooks-Gunn of Columbia University Teachers College, and Sheila Kamerman and Alfred Kahn of Columbia's School of Social Work. The works of authors Michael Katz, Frances Fox Piven, Richard Cloward, Lawrence Mead, Michael Harrington, and many more provided invaluable analysis. Journalists Alex Kotlowitz, Leon Dash, Jonathan Kozol, and Nicholas Lemann showed me by example how this book-writing stuff is done.

Dean Tom Goldstein and the entire faculty of Columbia University's Journalism School released me from teaching duties for a semester to work on this book. A special thanks to my colleagues Sam Freedman, Michael Shapiro, and Sig Gissler, for offering invaluable advice, and to James Carey for helping secure a Freedom Forum Professor's Publishing grant to complete the Moldova part of the book.

Friends and students contributed more than they know. Wayne Barrett, my reporting mentor in the early *Village Voice* days, encouraged me to pursue this book and continued to support the project by setting loose his army of interns to photocopy documents and background materials. Annette Fuentes filled in as my teaching adjunct when I had interviews to pursue. She and our mutual friend and colleague Maria Laurino offered astute insights and critiques of the manuscript along the way. Rob Polner provided insight and information during the book's inception. Two researchers helped in the early stages of the book: Adam Fifield did preliminary interviews and Shula Neuman started the research rolling.

I am grateful to have a wise and straight-shooting agent in Barney Karpfinger, who took a chance on a first-timer. Editor Doris Cooper at William Morrow originally signed the book and Henry Ferris took over all the hard work after Doris left for another house.

My children, Halley and Stefan Bondy, exhibited an ever-so-slight curiosity about what I was doing tapping away on the keyboard in the basement for months on end.

And most important, my mom and dad, who raised me to work hard and fly right, critiqued early versions and were far too kind.

Thank you.

Index